Vedic Geometry Course

Published by :
Lotus Press Publishers & Distributors

Vedic Geometry Course

Dr. S.K. Kapoor

Lotus PRESS

4735/22, Prakash Deep Building
Ansari Road, Darya Ganj,
New Delhi - 110002

Lotus Press Publishers & Distributors
Unit No.220, 2nd Floor, 4735/22, Prakash Deep Building,
Ansari Road, Daryaganj, New Delhi- 110002
Ph:- 41325510, 23280047, 098118-38000
Email : lotuspress1984@gmail.com
Visit us : www.lotuspress.co.in

Vedic Geometry Course

ISBN: 81-8382-047-6

Printed & Published by : **Lotus Press Publishers & Distributors,** New Delhi- 2

INNER VOICE FOREWORD

— FROM A SAINT

Hear your inner voice and be free from the outside limitations. Within is the seat of the Lord and it is here where we have to be face to face with our selves. This origin source reservoir is to be tapped, and this appears to be the urge and intensity of present attempt to reach at 'Vedic mathematics basics' in five steps. These five steps would prepare for the phase and stage to be in communication with the SELF.

The SELF is TRANSCENDENTAL. The seat of the SELF within *Triloki* is the Creator's space. Self established at the seat of Creator's space ensures carrying the Being along the transcendental carriers. It is the bliss of this ensured supply of transcendental carriers which intensifies the urge of the transcending mind to glimpse the inner folds of the transcendental worlds manifesting within Creator's space.

The whole efforts and beginning for it, as such is to be to attain transition from Triloki manifesting as outside mundane macro state world as first state of consciousness for the intelligence to the next transcendental micro state world of the intelligence lively within Creator's space as origin source reservoir of Triloki. Our ghan / body cube, as such, is to be chased for its origin source reservoir within it lively as Chatar mukhi / four head Creator.

Really blissful is the learning path of this chase range. It is transcendentally blessed. Let sadkhas permit their transcending mind to fulfill with bliss of the transcendental worlds / Panch Mukhi / five head lord sustaining the Creator's space.

Lively within
cavity of heart:
25-11-2005

In search of ***JYOTI***
manifesting as **NAD**

■

PREFACE

Urge to know ancient wisdom expressed as Vedic knowledge inherently imples acceptance for chase of it the way these systems permit. This as such would mean that the sadkhas for the chase of Vedic systems have to proceed along the established paths of *Sankhiya Nishshta* and *Yoga nishta* which not only run parallel to each but are also complementary and supplementary of each other.

The sankhiya nishtha, on its ultimate analysis amounts to availing the artifices of numbers, for coordination arrangement of organisation format, while on the other hand, the Yoga nishta avails the dimensional formats, for regulation of transcendence and ascendance of mind through the manifested creations. The **TRANS**, also means, the number ahead, and it is this way, the transcending mind ultimately avails the artifices of numbers to sequentially glimpse the ultimate core of the transcendental worlds.

The par-excellence feature of Vedic systems is that these approach the whole range of knowledge as a single Discipline, and as such, on ultimate analysis the organisation of Discipline of knowledge avails the dimensional frames parallel to the artifices of numbers. This, this way, shall be taking us to the artifices of numbers, and the dimensional frames as the source processing tools in terms of which even the self referral feature of transcending mind chasing its own transcending path can also be worked out intellectually.

It is this feature which deserves to be chased for the basics, and here in the present attempt, the urge being to learn and teach Vedic mathematics on its basics, as such, at this phase and stage of exposing young minds during Middle school learning, The focus is to be upon the centre of the cube / origin of 3-space being of the format of hyper cube-4 and as such seat of creator's space.

This focus naturally aims to eye the goals of attainments for the young minds for perfection of their intelligence by having transition from the set up of the Cube to transcend ahead for glimpsing the set up of hyper cube-4 along with its transcendental base of the format of hyper cube-5.

Though, the learning all about hyper cubes 4 and 5 is to be the content focus of high school classes, and ahead there to would be the subject focus of the formats of hyper cubes-6 and 7 during higher secondary classes, but the initiations for the foundations of appropriate mental state for those formats are to be ensured during schooling of middle school stage.

This way the learning range of this phase and stage of schooling during classes 5 to 8 is to be centred around the set up of the cube with focus upon its centre /

origin of 3-space as seat of creator's space (4-space) as of format of hyper cube -4. Accordingly the teaching and learning of this range of schooling is phased as five books namely, book-1The teaching of Vedic mathematics, book-2 Learning Vedic mathematics on first principles, book-3 Vedic mathematics basics, book-4 Vedic mathematics skills and book-5 Vedic geometry course, for perfection of intelligence.

These books, in their integrated approach of chase of the range for set up of the cube / 3-space with focus upon its center / origin of 3-space, as source reservoir for the features of the set up, is meant for exposure of young minds to acquire perfection of intelligence while learning basis of basics on first principles along the concrete format of a cube.

The first principles for approaching basis of basics of Vedic mathematics on solid formats are being supplied by the Ganita Sutras, particularly as the ordering principle of Ganita Sutra-1 which makes it as the source Sutra, and as such 'cube' and 'Ganita Sutra-1' are to ever remain within the eye sight of the learners of this phase and stage of schooling.

The aim and goal of learning being the chase of basis of basics on first principles to ensure perfection of intelligence for the young minds, as such heavy responsibiliy lies upon the Vedic mathematics teachers and parents to make the learning really blissful throughout the duration of chase.

The whole range of enlighened souls begining with Maharishi Valmiki and Maharishi Ved Vyas to Maharishi Dayanand and Swami Bharti Krsna TirthaJi Maharaj to Shri Pad Baba and Maharishi Mahesh Yogi, this learning path is being made blissfully lively, and the enlightened teachers are to continue the linege.

With these inspired background, present humble attempt has been made to reach at these books and it is hoped that these may provide required exposure for the young minds with the help of the enlightened teachers, who shall be making their instructions free of technicalities and the knowlede content to flow blissfully within the head s and hearts of the students.

Mr. Kenneth R. Williams and Mr. R. P. Cosmic Kapoor deserve my heartful thanks for remaining intellectually with me during these days of my reaching at these books. Mr. Rohtash Kapoor deserves full appreciation for conducting Vedic geometry course.

Mr. Deepak Girdhar deserves all appreciation for his nice graphics and the computer work.

Credit for nice publication goes to Lotus Press.

The inner voice forward from a saint gives me the satisfaction and confidence as that these efforts may be fruitful for the cause of basics to be brought home to the young minds.

Vedic home, **Author**
Rohini, Delhi
25-11-2005

■

CONTENTS

FIRST FEATURE
FOUR SPACE REALITY

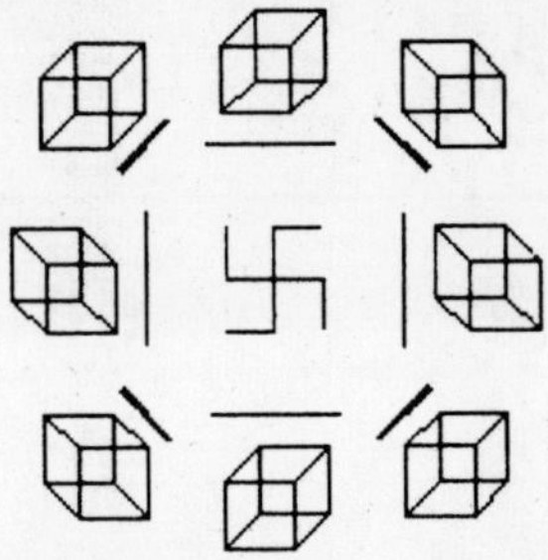

Versions of Hyper cube-4

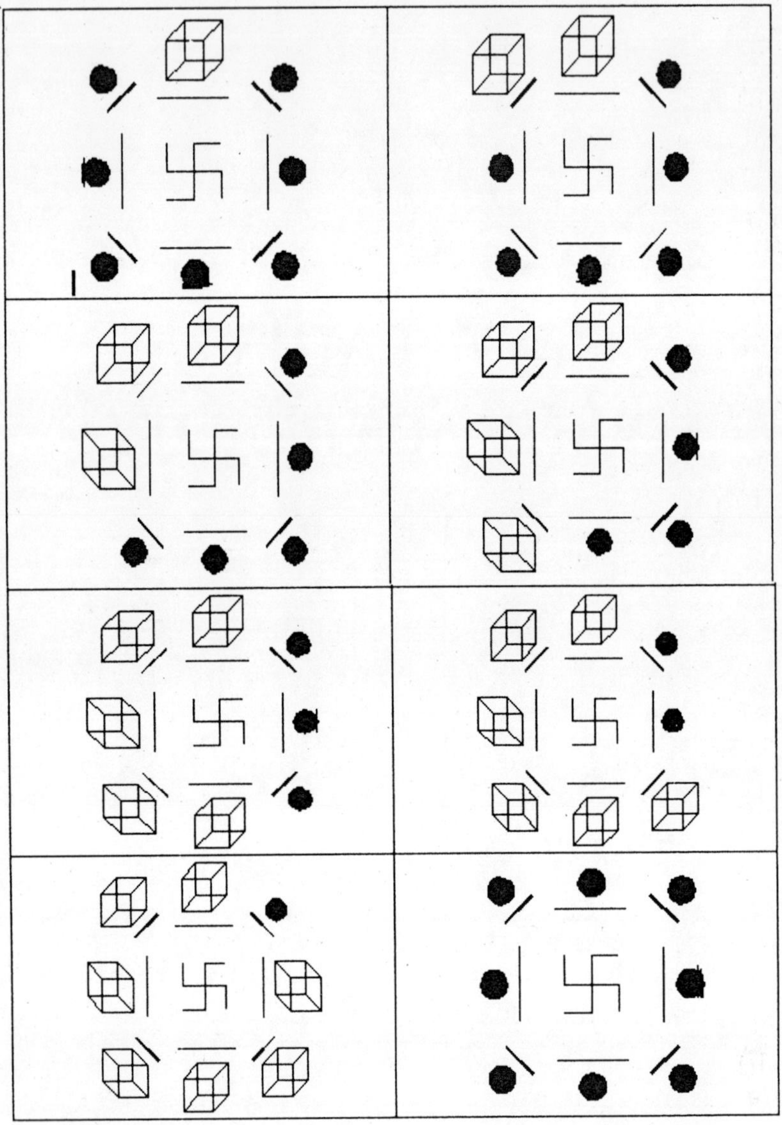

ASPECT-1.1
INTRODUCTION

Vedic systems accept two fold processing, designated and known as, ***sankhiya* system and *yoga* system**. Both are complementary and supplementary of each other. These also run parallel to each other. End fruit of processing plan either path is of same order and bliss. Therefore, either of it may be followed, or even as both being, complementary and supplementary of each other, so both may be availed simultaneously.

Sankhiya systems, on their ultimate analysis, emerge to be availing **artifices of numbers**, presuming the existence and availability of dimensional formats for the artifices. On the other hand, ***yoga* systems**, on their ultimate analysis emerge to be availing **dimensional formats**, presuming the existence and availability of artifices for the dimensional formats.

These characteristic features and presumptions of these processing systems make them to be complementary and supplementary of each other as well as these to be of parallel progressions and identical end fruits.

For the chase of the Vedic geometric feature of Real 4-space, we may avail:

1) Artifice 4, and
2) Space accepting dimensional frame of four dimensions

REAL 4-SPACE

Chase of this feature may first be as to reaching at the representative regular body of this space availing the characteristics of the format, manifesting as **IDOL** of **Lord Brahma**.

Designating this body as **hyper cube-4**, it as such is to be located as **fourth member** of the sequence of representative regular bodies of dimensional spaces. It would mean to reach at hyper cube-4 existing in continuity of the geometric set ups of interval, square and cube being the representative regular bodies of 1, 2 and 3-space respectively.

The **second approach** for location of Real 4-space and for reaching at the geometric set-ups of hyper cube 4, is to be **to reach at center of cube / origin of 3-space**. It would mean to have a cut of a cube into eight sub cubes parallel to a cut of three space within three dimensional frame into eight octants.

The **third approach** for comprehension of existence of Real 4-space is to be to comprehend the **moving tracks of points, lines, surfaces and solids** and to the melt mental block about the space standing finally blocked as tracks of moving surfaces.

The **tracks of moving surfaces leading to 3-space**, in no way can be taken as that the tracks of moving solids would stand restricted up-till the tracks of moving surfaces.

The **fourth way to approach** about the existence of Real 4-space would be to transcend through the macro state of physical objects and to reach the micro state of the entities. This would be reminding us as to how the human mind has been successful to fix the atomic world.

Fifth way to approach this phenomenon of existence at micro state is to permit the transcending mind to transcend through the mundane worlds of **waking state** of consciousness and to attain the next / **dream state** of consciousness as to the existence at the **micro state of subtle state body**.

■

For Further Attention

The origin base of 4-space is 5-space. Its symbolic notation is as follows:

■

ASPECT-1.2
EXISTENCE OF HIGHER SPACES

1. We can view
   (a) interval (line) / 1-space as a track of a moving point,
   (b) surface (square) / 2-space as a track of a moving interval / line / 1-space, and,
   (c) solid space (cube) / 3-space as a track of a moving surface / square / 2-space.

Interval	*Square*	*Cube*	*Hypercube-4*
		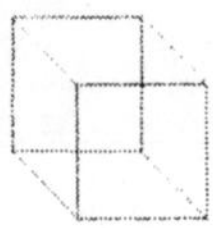	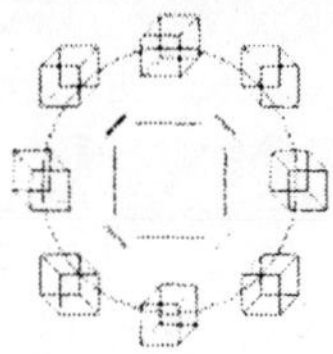

2. From here we can pose a question to ourselves as to what would be the track of a moving solid / cube / 3-space?
3. The natural answer is 4-space / hypercube-4 / hyper solid domain.

4. We are having free movements of solids. Even when we move, we move as a solid block, so we always create 4-space for our free motion.
5. So our space in which we move freely is not a 3-space. It is 4-space and higher spaces all compactified.
6. With the motion of hypercube-4 / 4-space, we shall be having 5-space for us and in general with the motion of hypercube-n / n-space, we shall be having (n+1) space.
7. As such the existence of higher spaces is one concept for whose comprehension and chase one shall perfect one's intelligence beginning with the motions of point, line, square and cube.
8. Once one has complete comprehension of existence of 4-Space with hyper cube-4 as its representative regular body with domain boundary ratio as $A^4:8B^3$, one shall proceed further for comprehension of 5-Space and its representative regular body.
9. Once one has complete comprehension of existence of 5-Space with hyper cube-5 as its representative regular body with domain boundary ratio as $A^5:10B^4$, one shall proceed further for comprehension of 6-Space and its representative regular body.
10. Initially, for perfection of intelligence, one shall sequentially go on chasing dimensional spaces 1 to 7 and their representative regular bodies. The subsequent chase of higher spaces shall be taken up only after one has perfected one's intelligence in respect of 1 to 7 spaces.

■

ASPECT-1.3
OUTWARD AND INWARD EXPANSIONS

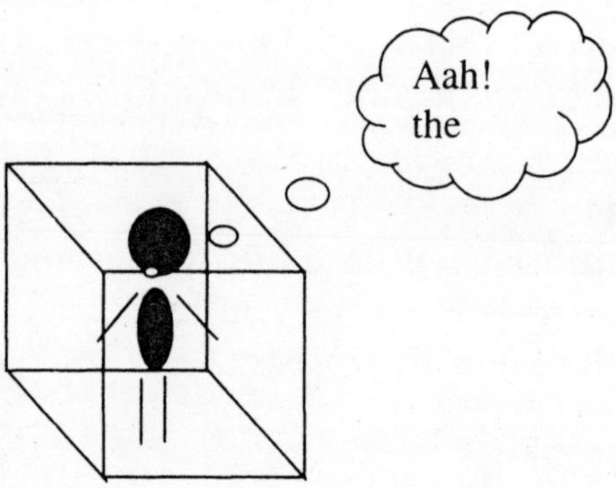

1. 4-space as a track of moving solid (cube), helps us to pause at the static position of a cube as within 4-space.
2. As the cube is wrapped within surfaces, so we can comprehend the situation as that 4-space is enveloping 3-space in terms of surfaces (2-space).
3. The surface plate of a cube when would be stripped off would fall within 4-space.
4. We can see conditioning and structuring of 4-space in terms of the surfaces (2-space).

5. This is 2-space playing the role of dimension of 4-space.
6. Cube is so structured that it at its center permits split up as 8 cubes parallel to the Cartesian cut of 3-space of 8 octants.
7. We can chase enveloping of the center of a cube as 8 cubes coming closer and closer.
8. This is as if the space at the center of cube / origin of Cartesian dimensional frame of 3-space / origin of 3-space is admitting wrapping within solid boundary of 8 components; a 4-space setup wrapped within solid boundary of 8 components in unison with the domain-boundary ratio $A^4:8A^3$.
9. With this we have outward expansion taking us from 3-space to 4-space with surface (2-space) at the boundary of the cube (3-space) structuring 4-space for us and simultaneously at origin of the 3-space / center of cube there being a seat of a 4-space which accepts 3-space domain at the boundary as a setup of 8 solid components as an inward expansion from 3-space to 4-space.
10. The phenomenon of simultaneous outward and inward expansion from 3-space to 4-space is a unique phenomenon and a beautiful chase of Vedic Geometry.
11. The diagram at the outset of this lesson above and the following diagram, when chased together as depictions of situations as to how, as is the situation of diagram-1 above, we first condition ourselves by accepting ourselves as if we are within 3-space and then starting working out linear orders and getting sealed at 3-space.

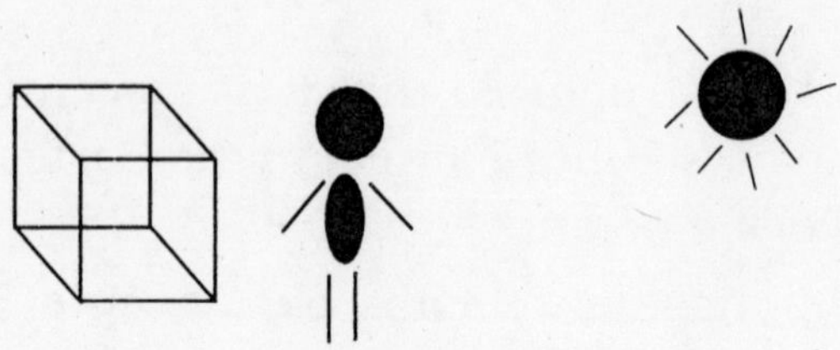

12. If we remind ourselves that we can come out of it and actually come out of it and then the change from the state of first diagram to the second diagram would amount to convincing ourselves that we are in a 4-Space. So perfected intelligence, further shall be availed for taking our selves out of initial position of being within hyper cube-4 and then coming out of it to be in 5-Space and this chase may be continued farther and farther:

COMPREHENSION EXPRESSIONS

1. Express the way the four space reality is coming within your comprehension.
2. Express your comprehension of representative regular body of 4-space existing in continuity of interval, square and cube as fourth member of the sequence of representative regular bodies of dimensional spaces.
3. Permit your transcending mind to experience this higher dimensional reality.
4. Share your experiences about the higher spaces reality.
5. Compare your experiential and intellectual comprehension of 4-space and its representative regular body (hyper cube-4).
6. Evaluate yourself about your comprehension and share it.

■

For further attention

Symbolic notation for hyper cube-5, the representative regular body of 5-space is as follows:

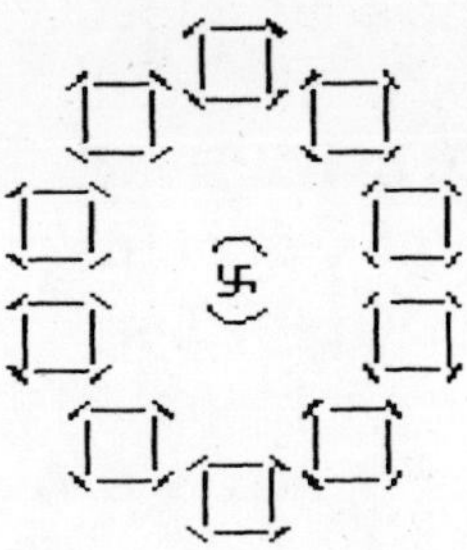

■

ASPECT-1.4
FIRST APPROACH TO THE FEATURE

To reach at the format, manifesting as **idol** of **Lord Brahma**.

COMPREHENSION EXPRESSION EXERCISES

Express the way the four space reality is coming within your comprehension.

SCRIPTURES DESCRIPTION OF IDOL OF LORD BRAHMA

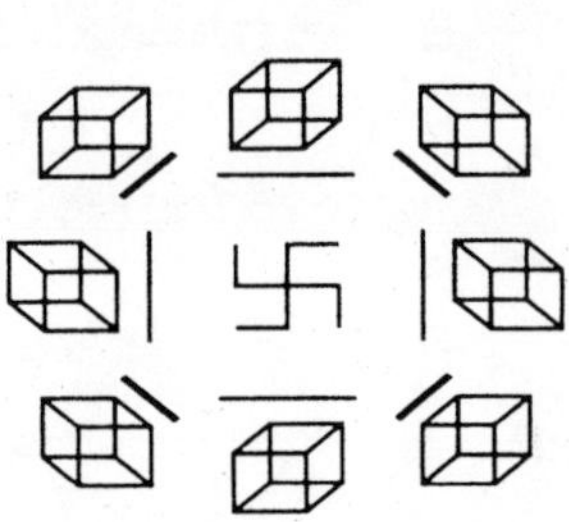

Scriptures describe Lord Brahma, Creator the supreme, as four head Lord sitting majestically upon his lotus seat of eight petals. He is equipped with a pair of eyes in his each head. Within cavity of his Heart is the transcendental seat of his Lord, Lord Shiv. IDOL of Lord Brahma depict the moments of bliss of Creator, the supreme, mediating upon his Lord, Lord Shiv, five head Lord of transcendental worlds. With grace of Lord Shiv,

Lord Brahma goes transcendental, multiplies ten folds, as ten brahmas enveloping Shiv Lok.

This description of IDOL of Lord Brahma may be chased parallel to the structural set up of hyper cube 4 of spatial order with its domain boundary ratio being $A^4 : 8B^3$

The following comparative table may help to comprehend the parallel features of IDOL of Lord Brahma and hyper cube 4.

Sr.	Feature	Hyper cube 4	IDOL
1	Dimensions	4	Four heads
2	Dimensional order	Spatial	Pair of eyes in each head
3	Boundary components	8	Lotus of 8 petals
4	Origin	5-space	Seat of Lord Shiv within cavity of heart of Lord Brahma

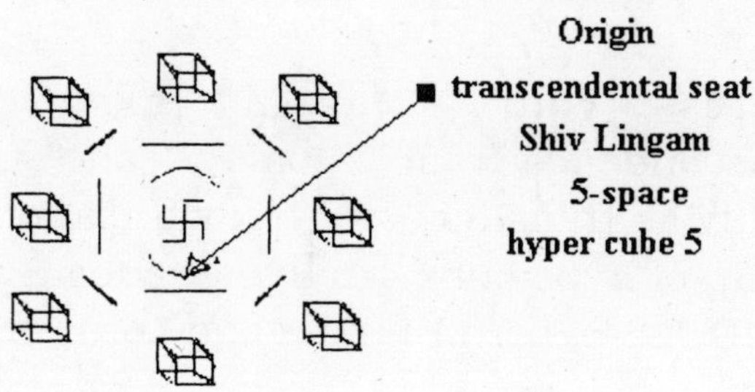

The boundary of the transcendental worlds is constituted by ten Creator's spaces ($A^5 : 10B^4$).

EXERCISES STEPS

1. Sit comfortably
2. Permit the mind to transcend.

3. Float the thought of 4-space.
4. Permit the mind to chase 4-space parallel to hyper cube 4 as well as parallel to IDOL of Lord Brahma.
5. Permit the mind to transcend time and again and to have bliss of Lord Brahma multiplying ten fold at boundary of transcendental worlds.
6. Between two sittings of trans, one shall have intellectual chase of hyper cube 4 in continuity of the set ups of interval, square and cube accepting domain boundary ratio as A^N: 2N B^{N-1}, N=1, 2, 3, 4.
7. Express your self about your comprehension of Reality of 4-space.
8. Share your evaluation of your comprehension of Reality of 4-space.
9. Enlist your sequential progression steps of this chase and comprehensions of 4-space Reality.
10. Enlist and share your blissful moments, as well as the urge to know further about this transcendental phenomenon of the origin of 4-space.
11. Express your glimpsing experiences of transcendence from 4-space Reality to 5-space Reality as transition and transformation from the set up of hype cube 4 to hyper cube 5 parallel to the IDOLS of Lord Brahma and Lord Shiv.
12. Express your glimpsing experiences of hyper cube 5 availing the transcendental format of IDOL of Lord Shiv, five Head lord equipped with three eyes in his each head. He has ten beautiful arms and presides over the transcendental worlds.

■

ASPECT-1.5
SECOND APPROACH TO THE FEATURE

To reach at center of cube / origin of 3-space. It would mean to have a cut of a cube into eight sub cubes parallel to a cut of three space within three dimensional frame into eight octants.

COMPREHENSION EXPRESSION EXERCISES

Express the way the four space reality is coming within your comprehension.

CUBE AS A SET OF EIGHT SUB CUBES

1. One may take a soap cake and a knife.
2. Cut the soap cake to make it 'two sub cakes'.
3. It be taken as a cut along one of the three 'axes'.
4. A further cut along another axis shall be cutting the cake / a pair of sub cakes into four 'sub sub cakes'.
5. A still further cut along third axes shall be cutting cube into a set of eight 'sub sub sub – cakes'.
6. This exercise would bring us, face to face with the set up of a cube as of 'sub cubes', eight in number.

7. The focus is upon the feature as that cube accepts synthetic set up in terms of eight sub cubes.

REPEAT IN REVERSE STEPS

8. The exercise may be repeated but in reverse steps, of building a cube with the help of another set of eight cubes.
9. In terms of above available eight 'sub sub cakes', as a starting point, the soap cake of original pre cut state can be reached.
10. This would help us have a comprehensive view of the composition of a cube as a set of eight cubes.
11. This also shall be helping us have comprehension of the specific features of the center of the cube as a seat for the one of the corner points of each of the eight cubes / sub sub sub cakes of the soap.
12. It may be viewed as that the center of the cube is of unique features which distinguish it apart from all other points, as much as that center of the cube may be viewed as enveloped within eight cubes.

FEATURES OF CENTER OF CUBE

13. This feature of the center as being enveloped within eight cubes / solids deserves to be chased further.
14. It may be chased as a set up existing in continuity of the geometric set ups of interval, square and cube, as much as that interval has a pair of boundary points, while square has a double pair of boundary lines, and cube has a triple pair of boundary surfaces. A step ahead is going to be a four pairs of boundary solids set up of hyper cube

4, being the fourth member of the sequence of representative regular bodies of dimensional spaces.

15. This takes us to the set up for the center of the cube as a set up of hyper cube 4 at its 0 values for the formulation $A^4:8B^3$, for A=0 and B=0; a set up of the state of 4-space and its representative regular body, in the role of origin of 3-space / center of cube.
16. It is this set up and set of features of center of the cube as of hyper cube 4 format which may help comprehend as to how the cube is when sequentially reduced to limit, as zero value, all of its corner points, edges and surfaces, as well as volume as a set up of eight sub cubes, simultaneously fall down at the center and the whole structure of the cube revives back uniformly with sequentially increasing cubes.

EXERCISES

1. It would be a blissful exercise to chase sequential decrease of the cube to limit zero taking to the center of the cube.
2. It would be a further blissful exercise to chase sequential increase of cube to 'unit cube' beginning from 'zero value cube'.
3. Express the way the four space reality is coming within your comprehension with its chase as a set up features of center of the cube has 0 value state of hyper cube 4 as representative regular body of 4-space.
4. Evaluate your comprehension and share it with us.

■

ASPECT-1.6
THIRD APPROACH TO THE FEATURE

To comprehend the **moving tracks of points, lines, surfaces and solids** and to melt mental blocks about the space standing finally blocked as tracks of moving surfaces.

COMPREHENSION EXPRESSION EXERCISES

Express the way the four space reality is coming within your comprehension.

PERMIT THE TRANSCENDING MIND TO CHASE TRACKS OF MOVING BODIES

PERMIT THE MIND TO TRANSCEND

1. Sit comfortably.
2. Permit the mind to transcend.
3. Let the mind chase a track of a moving point.

PERMIT THE MIND TO GLIMPSE THE TRACK OF A MOVING POINT

4. It would be blissful to glimpse the track of a moving point manifesting a line, as a straight line, as a

curve within a surface, a spiral within 3-space and hyper spirals within 4-space.

PERMIT THE MIND TO GLIMPSE THE TRACK OF A MOVING LINE

5. It would further be blissful to glimpse, in continuity the track of a moving line manifesting a surface, as a plane, as a spatial spiral with 3-space, and a solid spiral within 4-space.

PERMIT THE MIND TO GLIMPSE THE TRACK OF A MOVING PLANE

6. It would be further blissful to glimpse, in continuity, the track of a moving plane manifesting a solid, as a cube, as a hyper solid within 4-space.

PERMIT THE MIND TO GLIMPSE THE TRACK OF A MOVING SOLID

7. It would further be a blissful to glimpse, in continuity the track of a moving solid manifesting as hyper solid fold within 4-space.

PERMIT THE MIND TO GLIMPSE THE TRACK OF A MOVING HYPER SOLID

8. It would be a transcendental experience to glimpse the moving hyper solids manifesting within transcendental worlds as of transcendental features.

COMPREHENSION EXPRESSION EXERCISES

1. Express the way the four space reality is coming within your comprehension during chase of the tracks of moving bodies, particularly of moving solids, and also of moving hyper solids.

2. Evaluate your comprehension of Four Space Reality and share the same with us. ■

For Further attention: Versions of hyper cube-5

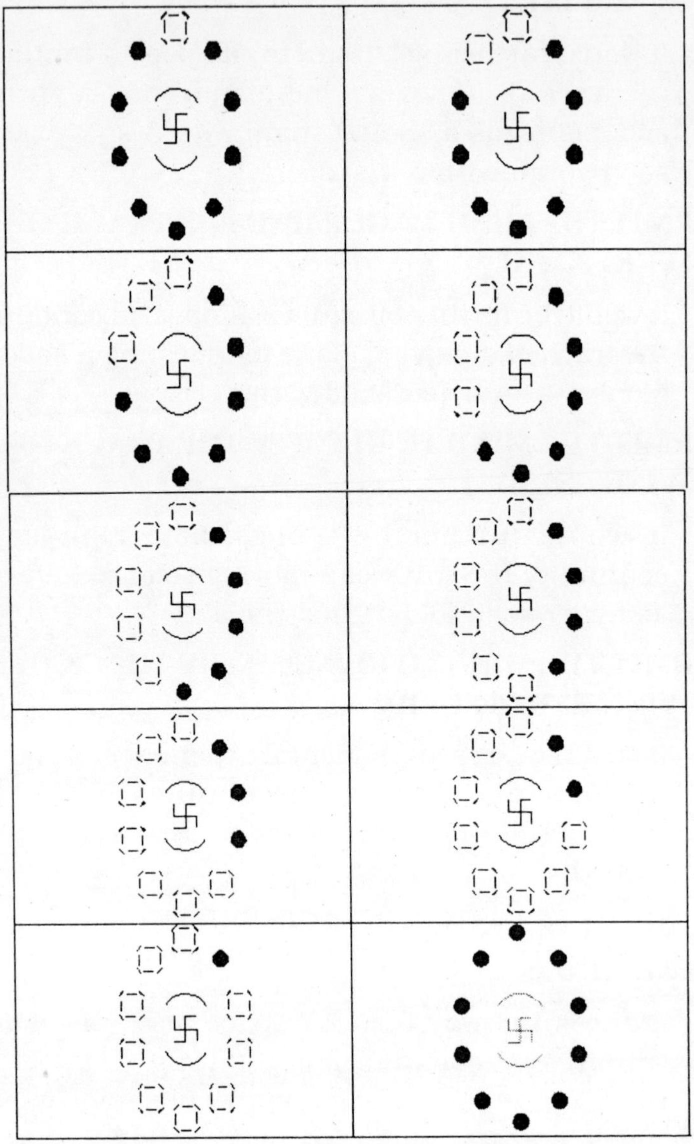

ASPECT-1.7
FOURTH APPROACH TO THE FEATURE

To transcend through the macro state of physical objects and to reach the microstate of the entities. This would be reminding us as to how the human mind has been successful to fix the atomic world.

2. COMPREHENSION EXPRESSION EXERCISES

Express the way the four space reality is coming within your comprehension.

MACRO STATE AND MICRO STATE

1. Transition from macro state to micro state may be chased as a transition from CUBE to HYPER CUBE 4.
2. It is the predominant expression of VOLUME of the cube which becomes the feature of MACRO STATE.
3. The CENTER of the CUBE becomes the feature of MICRO STATE.
4. The center of the cube is within cube at its zero value state but is of the format of hyper cube 4.

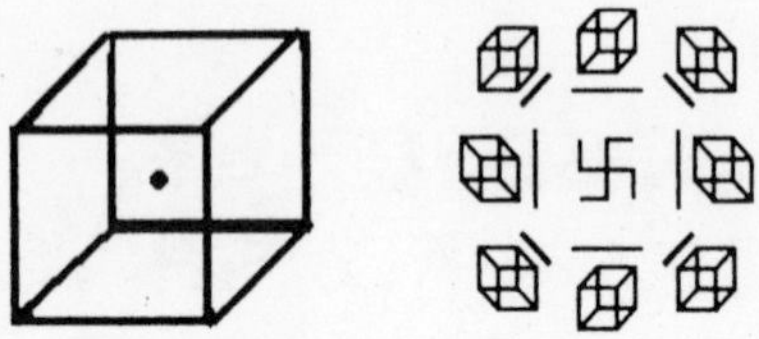

5. One may approach for this transition by sequentially reducing the CUBE of smaller and smaller volume and reach a phase and stage of limit of volume as limit of macro state.
6. Then, one may approach for this transition by sequentially reducing hyper cube 4 and reaching at the center of the cube as limit state of microstate.
7. One may take clue for macro state limit from the initial approach for access to 'atom'. It may be designated as initial macro state model of 'atom', as the smallest possible state within catch.
8. Then one may take clue for micro state model format for atom of electrons configurations.
9. Both the above approach attempts may be worked parallel to each other to have a view of micro state reality vis-à-vis the macro state reality.
10. The feature of the potentialities for the outer most orbit of the atom as to be of capacity to sustain at the most eight electrons, may be viewed parallel to the solid boundary of hyper cube 4 being of precisely eight components.
11. The plane permitting four directions and four angles, together as a frame of eight features, the cube permitting split up as of eight sub cubes and

the hyper cube 4 accepting eight solid components and also the outer most orbit of atom having sustaining capacity and potentiality for retaining eight electrons are the features which may help have firm comprehension of reality of 4-space in terms of hyper cube 4 as its representative regular body providing format for atoms, as well as the same marking its presence within cubes and squares.

12. One may permit ones transcending mind to chase these features of format of atoms, hyper cube 4, eight solid components boundary of hyper cube 4, split up of 3-space into eight octants and cube into eight sub cubes while geometric set ups of a square accepting a frame of eight features being 4 directions and 4 cones.
13. This experiential bliss, need be expressed and shared.

COMPREHENSION EXPRESSION EXERCISES

Express the way the four space reality is coming within your comprehension.

1. Evaluate your experiential bliss and reach at the level of your comprehension of 4-space reality and share it with us.

LINKING NOTE

I

1. FIFTH APPROACH TO THE FEATURE

To permit the transcending mind to transcend through the mundane worlds of **waking state** of consciousness and to attain the next / **dream state** of consciousness as to the existence at the **micro state of subtle state body**.

2. COMPREHENSION EXPRESSION EXERCISES

Express the way the four space reality is coming within your comprehension.

II

1. MATHEMATICS, SCIENCE AND TECHNOLOGY OF 4-SPACE

A Mathematics of 4-space

B Sciences of 4-space

C Technologies of 4-space

2. COMPREHENSION EXPRESSION EXERCISES

Express the way the four space reality is coming within your comprehension.

III

1. MATHEMATICS, SCIENCE AND TECHNOLOGY OF 4-SPACE

A Mathematics of 4-space

B Sciences of 4-space

C Technologies of 4-space

2. COMPREHENSION EXPRESSION EXERCISES

Express the way the four space reality is coming within your comprehension.

■

For further studies

Book Vedic geometry of the Author may be referred.

■

FEATURE-2

DISTINGUISHING FEATURES OF HYPER CUBE 4

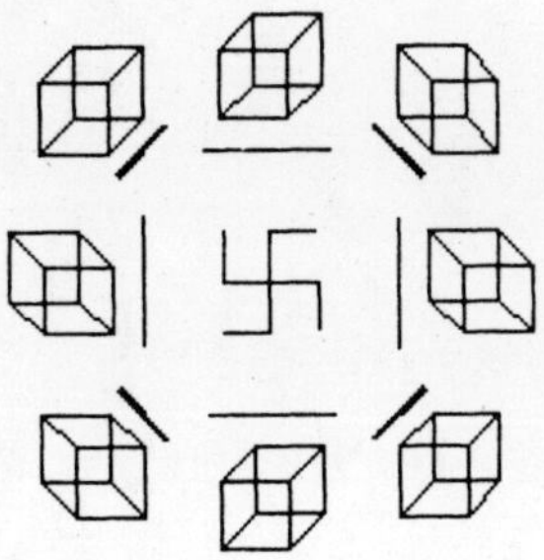

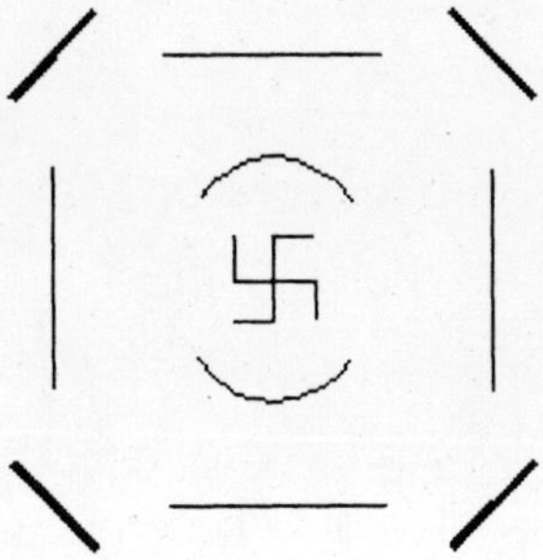

FEATURE-2

The **second feature** is about the distinguishing features of this space vis-à-vis our three space, whose chase is aimed to be had by comparative study of representative regular bodies of three space and four space respectively.

ASPECT 2.1
FIRST DISTINGUISHING FEATURE

ASPECT 2.1 A INTRODUCTORY

1. Each dimensional space has its distinct representative regular body.
2. The representative regular body of a dimensional space, formally may be defined as a dimensional body of the concerned space which does not distinguish between one or the other dimension of the space.
3. Illustratively interval is the representative regular body of 1-space, and square is the representative regular body of 2-space.
4. Cube is the representative regular body of 3-space as cube does not distinguish between one or the other dimension (axis) of 3-space.
5. There may be more than onc representative regular bodies of a space. Illustratively, cube as well as sphere are the representative regular bodies of 3-space.
6. However here we are proceeding ahead with cube as the representative regular body of 3-space.

7. Hyper cube 4 is a representative regular body of 4-space as it does not distinguish between either of its four dimensions.
8. Here, as second feature of the course, the comparative study of the distinguishing features of cube and hyper cube 4 are being taken up.
9. There are going to be a very large number of distinguishes features of cube and hyper cube 4 but here only three aspects of them are being focused, namely, firstly as to the different geometric entities prominently marking their presence in making of the set ups of cube and hyper cube 4, secondly as to the nature of the origins of 3-space and 4-space, and thirdly about the different roles which 3-space / cube and 4-space / hyper cube 4 shall be playing.
10. Of these, the first aspect itself is being touched only for three of its prominent characteristics, viz., (1) the number of dimensions, (2) nature of dimensions and (3) the geometric envelops with focus upon boundaries of cube and hyper cube 4
11. As far as the second aspect, concerning the origins of 3-space and 4-space is concerned, the same is being taken up only in its generality.
12. The third aspect as well is being touched for its three prominent characteristics, viz. 3-space and 4-space being (1) In the role of dimension, (2) In the role of boundary and (3) In the role of origin.
13. As such, these three aspects, mark out seven prominent features to be touched here during this week.

First feature of cube and hyper cube 4 as on today (22-9-2005)

14. The first prominent feature in terms of which the geometric set ups of cube and hyper cube 4 may be distinguished, the same is as per the number of dimensions which are responsible for these set ups of cube and hyper cube 4.

Second feature of cube and hyper cube 4 as on 23-9-2005

15. The second prominent feature in terms of which the geometric set ups of cube and hyper cube 4 may be distinguished, the same is as per the nature of the dimensions themselves. In other words, here the focus is upon the geometric set ups of dimensions / axes of 3-space as well as of 4-space.

Third feature of cube and hyper cube 4 as on 24-9-2005

16. The third prominent feature in terms of which the geometric set ups of cube and hyper cube 4 may be distinguished, the same is as per the nature of the geometric envelops of cube as well as hyper cube 4. In other words, here the focus is upon the geometric set ups of boundaries of cube as well as hyper cube 4.

Fourth feature of cube and hyper cube 4 as on 25-9-2005

17. The fourth prominent feature in terms of which the geometric set ups of cube and hyper cube 4 may be distinguished, the same is as per the nature of the origins of 3-space and 4-space.

Fifth feature of cube and hyper cube 4 as on 26-9-2005

18. The fifth prominent feature is about the role of 3-space as dimensions and the role of 4-space as dimensions.

Sixth feature of cube and hyper cube 4 as on 27-9-2005

19. The sixth prominent feature is about the role of 3-space as boundary and the role of 4-space as boundary.

Seventh feature of cube and hyper cube 4 as on 28-9-2005

20. The seventh prominent feature is about the role of 3-space as origin and the role of 4-space as origin.

ASPECT 2.1 A COMPARATIVE STUDY OF CUBE AND HYPER CUBE 4

cube	Hyper cube 4
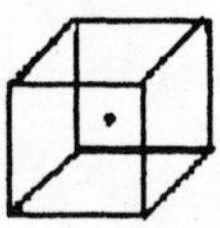	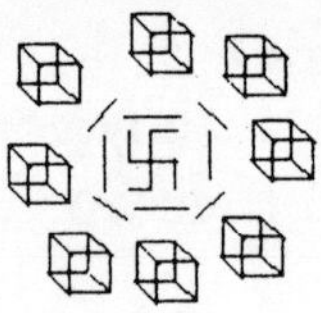

FIRST DISTINGUISHING FEATURE

The first prominent feature in terms of which the geometric set ups of cube and hyper cube 4 may be distinguished, the same is as per the number of dimensions which are responsible for these set ups of cube and hyper cube 4.

1. Space is better known by it's representative body.

Illustratively, 'interval supplies us knowledge' about '1-space', and likewise 'square supplies us knowledge' about 2-space.

2. Likewise, we may reach at prominent features of 3-space and 4-space, in terms of the comparative study of their representative regular bodies, namely 'cube' and 'hyper cube 4' respectively.
3. First prominent feature of dimensional spaces is the dimensional frame of a space.
4. The dimensional frame of a space may be formally defined as a set up of dimensions tagged with the origin.
5. Our well known, three dimensional frame, is a set up of three lines tagged with origin (common point) with the further restriction as that all the three lines not in a plane.
6. The said three lines are known as 'axes'. These are designated as three dimensions of 3-space.
7. It leads to a feature as that, in the context of 3-space, 3 lines, 3 axes and 3 dimensions are of same features.
8. These common features of lines, axes and dimensions, as such, lead to a conclusion as that 3-space is a linear order space as that it accepts linear dimensions, 3 in number.
9. The three dimensional frame, as such, does not permit any fourth line which would not be a co planer with either axis.
10. This way, a 3 dimensional space, may be formally taken as a closed set up within a 3 dimensional frame.

11. The Vedic systems, instead approach 3-space in terms of different set ups for the dimensional frames, prominent amongst them being designated as a measuring rod constituted by the representative regular bodies of 1, 2 and 3-spaces.

12. This dimensional frame / measuring rod, formally may be depicted, in more than one ways, and two of these ways, are, firstly a sathapatya measuring rod and secondly the format of Maheshwara Sutras, beginning with Maheshwara Sutra-1.

13. The Sathapatya measuring rod for 3-space is linear constituted set up of closed interval, square and a cube, together synthesized in that sequential progression order, as a synthetic measuring rod of progression order.

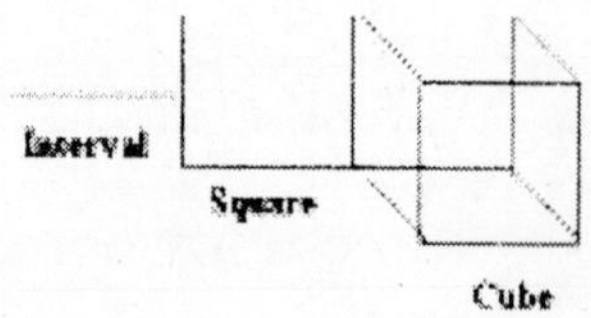

14. The Maheshwara Sutras avail this measuring rod as a dimensional frame for their organization format, beginning with first Maheshwara Sutra format as follows.

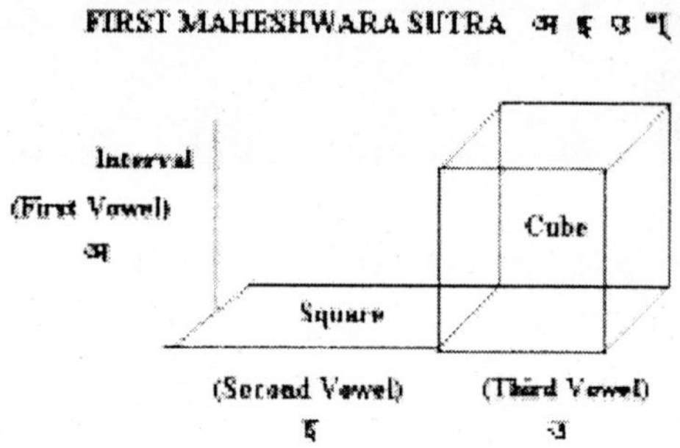

15. At this initial stage, one may avail the sathapatya measuring rods of 3-space and 4-space for

comparative study of cube and hyper cube 4 for reaching at distinguishing feature of 3-space and 4-space.

16. The sathapatya measuring rod of 3-space is constituted by representative regular bodies of 1, 2 and 3-space, and may be depicted and chased as follows.

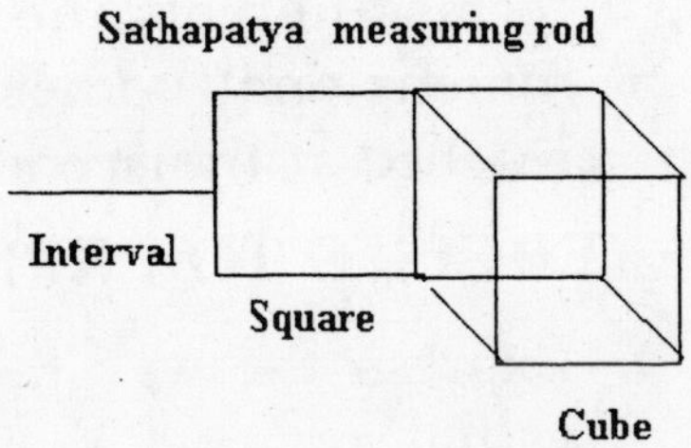

17. The sathapatya measuring rod of 4-space is constituted by representative regular bodies of 1, 2, 3 and 4-space and may be depicted and chased as follows.

18. One way to avail sathapatya measuring rod of 3-space is to dive deep in 3-space / cube with the help of measuring rod of 3-space. This in other worlds would mean that within the cube there exist sequential steps of interval, square and cube.

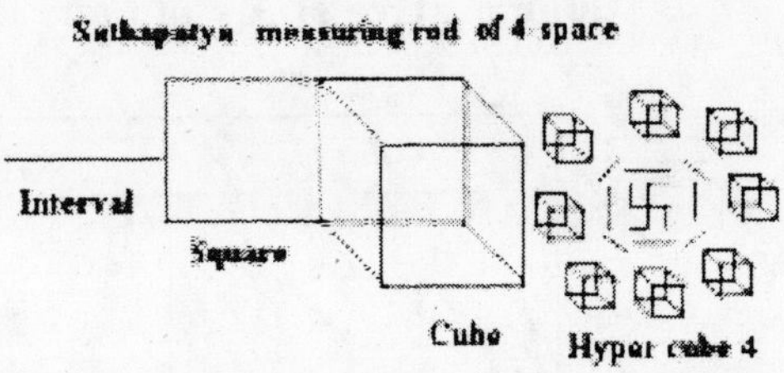

19. The parallel way to avail sathapatya measuring rod of 4-space is to dive deep in 4-space / hyper cube 4 with the help of measuring rod of 4-space. This in other worlds would mean that within the hyper cube 4 there exist sequential steps of interval, square and cube.

20. The sequential steps of sathapatya measuring rod are parallel to the dimensional frames of the concerned dimensional spaces, and as such, in terms thereof, the comparative study of the dimensional bodies and their spaces may be reached at transcending the blocking of 3-space.

EXERCISES TO EXPRESS AND SHARE

1. Express the way the 4-space reality is coming within your comprehension in terms of sathapatya measuring rods of 3-space and 4-space.
2. Evaluate your comprehensions of 4-space by comparative distinguishing features of sathapatya measuring rods of 3-space and 4-space.

2ND FEATURE

Comparative study of cube and hyper cube 4

cube	Hyper cube 4
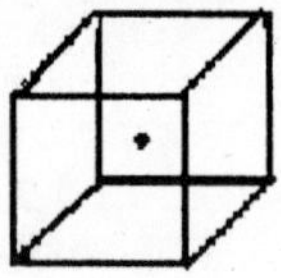	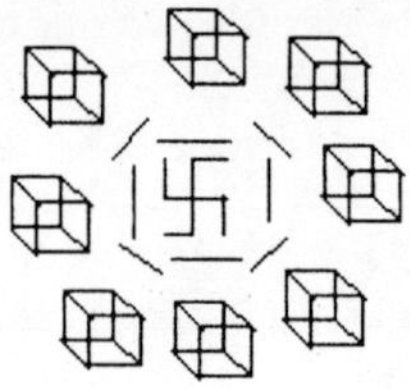

The second prominent feature in terms of which the geometric set ups of cube and hyper cube 4 may be distinguished, the same is as per the nature of the dimensions themselves. In other words, here the focus is upon the geometric set ups of dimensions / axes of 3-space as well as of 4-space.

ASPECT-2.2
GEOMETRIC SET UPS OF DIMENSIONS / AXES OF 3-SPACE AS WELL AS OF 4-SPACE

1. There are three dimensions / axes of three space.
2. The measuring rod for 3-space spreads out as a synthetic set up of 'interval, square and cube'.
3. All 'intervals, squares and cubes' are there in a CUBE itself.
4. The other prominent feature of set ups of 'interval, square and cube' is that interval and one axis of 3-space dimensional frame have parallel formats. Likewise square and a pair of axes of 3-space dimensional format as well have parallel formats. Further the cube and all the three axes of 3-space dimensional frame accept parallel formats.
5. Along, artifices of numbers, the three dimensional frame can be chased as artifice 3.
6. Also the artifice 3 and 3 dimensional frame can be chased as artifices '1, 2 and 3'.
7. The artifices '1, 2 and 3' constitute a unique set of numbers which are accepting no divisor other than '1' or the number itself.
8. These artifices, as such, may be designated as

primes with definition of prime being a number which accepts 'no divisor other than '1' or the number itself'.

9. The unique expression for this set of artifices is of the format 1 × 1, 1 × 2 and 1 × 3.
10. The artifice ahead, that is, 4 is a composite number, as its accepts in addition to 1 & 4, another divisor for it, namely '2'.
11. This, this way makes a departure from the set up of the artifices '1, 2 & 3', and hence from the 3 dimensional frame which is constituted by 3 dimensions / axes of '1-space / lines formats'.
12. The 4-space set up accepts 4 dimensions.
13. The measuring rod of 4-space is constituted by the representative regular bodies of 1 to 4 space, that is, by 'interval, square, cube and hyper cube 4'.
14. Parallel to it, along artifices of numbers, this chase can be in terms of quadruple of artifices (1, 2, 3, 4).
15. The insight about the nature and geometric set ups of dimensions / axes / dimensional order of hyper cube 4 / 4-space / Creator's space / of format of idol of Lord Brahma, the four head Lord, with each head equipped with a pair of eyes, may be had by chasing as follows.
    (i) Along artifice 4, as a composite artifice, also accepting '2' as a factor.
    (ii) Each head of Creator, the presiding deity of 4-space is equipped with a pair of eyes, which makes spatial order for each of the four heads.

(iii) The scriptures, and to quote one amongst them, Swateshwara Upnishad, following '1, 2, 3, 8' is attained liberation from the mundane worlds.

The quadruple of artifices '1, 2, 3, 8' are of the format (1 x 1, 2 x1, 3 x1, 4 x 2). This format leads to the message as that along the first three steps of a measuring rod, one is to go by the linear order and at its fourth step is to be followed the spatial order.

16. As such the distinguishing feature of 3-space / cube and 4-space / hyper cube 4 lies in the nature of their dimensions, which in the case of 3-space is of linear order and the same in case of 4-space is of spatial order. Which in other worlds means as that 1-space plays the role of dimension of 3-space while 2-space plays the role of dimension of 4-space.

RECAPITULATION AND SUM UP

Distinguishing features of 3-space and 4-space

Sr. no.	3-space	4-space
First	3 dimensions	4 dimensions
Second	Linear dimensions	Spatial dimensions

II EXERCISES TO EXPRESS AND SHARE

7. Express the way the 4-space reality is coming within your comprehension in terms of the nature of the dimensions of 3-space and 4-space.
8. Evaluate your comprehensions of 4-space by comparative distinguishing feature of the nature of the dimensions of 3-space and 4-space.

ASPECT-2.3
THIRD DISTINGUISHING FEATURE

1. One shall observe the geometric set up of a cube.
2. One shall be observing as that 'Volume / 3-space content lump' is contained within a geometric envelope stitched as '6 surfaces, 12 edges and 8 corner points'.
3. This geometric envelope can be gone into, in terms of a sathapatya measuring rod constituted by 'interval, square and a cube'.
4. Now one shall observe the geometric set up of hyper cube 4.
5. One shall be prominently observing the distinguishing feature of the geometric envelope of hyper cube 4 from that of the geometric envelope of cube, as that while cube requires stitching of 'six squares being surface plates', on the contrary, the requirement of hyper cube 4 is that of '8 cubes being boundary components'.
6. One can go into hyper cube 4 as 4-space domain in terms of sathapatya measuring rod constituted by 'interval, square, cube and hyper cube 4'.

7. To transcend through the geometric envelope of a cube stitched in terms of 'squares / surface plates' and to dive deep into the 3-space domain in terms of the measuring rod constituted by 'interval, square and cube / representative regular bodies of 1-space, 2-space and 3-space' would focus upon the 'boundary surface plates' which are six in number.
8. To transcend through the geometric envelope of hyper cube 4 stitched in terms of 'cubes / solid boundary components' and to dive deep into the 4-space domain in terms of the measuring rod constituted by 'interval, square, cube, and hyper cube 4 / representative regular bodies of 1-space, 2-space, 3-space and 4-space' would focus upon the 'solid boundary components', which are eight in number.
9. As such, the distinguishing feature of 'cube and hyper cube 4' comes to focus as set ups of their boundaries.
10. The boundary of cube being constituted by 'squares / surface plates', leads to focus as that '2-space' is in the role of boundary of '3-space'.
11. The boundary of hyper cube 4 being constituted by 'cubes / solid boundary components', leads to focus as that '3-space' is in the role of boundary of 4-space.
12. With it, it may be concluded as third distinguishing feature of 3-space and 4-space in terms of the comparative study of their representative regular bodies, that is, cube and hyper cube 4, as that while 3-space accepts 2-space as its boundary, on

the other hand, 4-space accepts 3-space as its boundary.

13. In other worlds, in terms of this distinguishing feature, 3-space and 4-space get coordinated for their co-existence as 'domain fold and boundary fold', as much as that 3-space / cube is to play the role of boundary fold of 4-space / hyper cube 4 as domain. The same also may be viewed as that 4-space / hyper cube 4 is to play the role of domain fold for 3-space / cube being in the role of boundary fold.
14. One shall experience this Reality of co-existence of more than one spaces in same geometric set up. For it, one may, first observe the set up of a cube and comprehend the simultaneous existence of 2-space and 3-space in the set up of a cube. One shall be observing as that 2-space, as surface plates, is manifesting as a boundary, while 3-space, solid content, is manifesting as a domain / volume of the cube.
15. When one shall be shifting to hyper cube 4, one shall be observing as that in this set up, 3-space is manifesting as a boundary while 4-space is manifesting as a domain of hyper cube 4.

RECAPITULATION AND SUM UP

Distinguishing features of 3-space and 4-space

Sr. no.	3-space	4-space
First	3 dimensions	4 dimensions
Second	Linear dimensions	Spatial dimensions
Third	Spatial boundary	Solid boundary

EXERCISES TO EXPRESS AND SHARE

9. Express the way the 4-space reality is coming within your comprehension in terms of the nature of boundaries of 3-space and 4-space.
10. Evaluate your comprehensions of 4-space by comparative distinguishing feature of the nature of the boundaries of 3-space and 4-space.

■

Chase organisation formats to attain transition

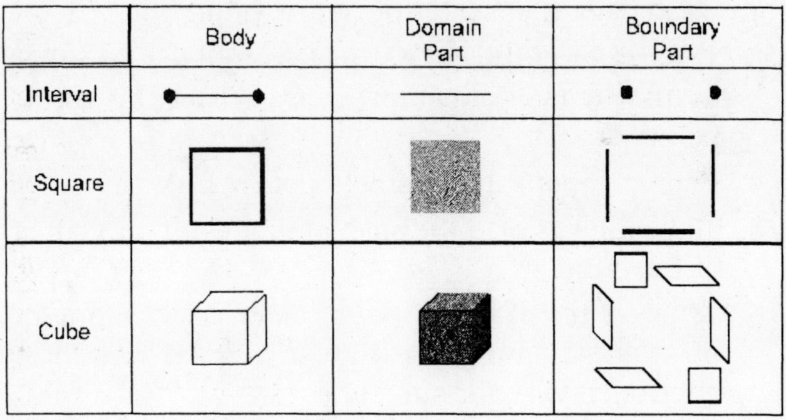

■

ASPECT-2.4
ORIGINS OF 3-SPACE AND 4-SPACE

1. Let us recapitulate the way, we had approached the center of a cube and origin of 3-space during first feature chase of 4-space Reality.
2. Let us again observe the set up of a cube.
3. The center of the cube is a class in itself, as much as that it is equidistant from each of the corner points.
4. Center constitutes a single point class.
5. The center is also the sole joint of eight sub cubes of the cube.
6. The center of the cube, as such is of a unique position, as much as that, it may be viewed as enveloped by eight sub cubes.
7. Parallel to center of the cube is the geometric phenomenon of eight octants cut of a 3-space.
8. This geometric phenomenon of cut of 3-space into eight octants is of the form and nature of features of origin of 3-space being enveloped by eight octants / solids.
9. This geometric phenomenon is of the form and

nature of the features of a set up of a hyper cube 4 as 4-space enveloped within 3-space as a solid boundary of eight octants.

10. This as such makes the origin of 3-space as being of the format of a 4-space and the center of a cube as being of the format of hyper cube 4.
11. The center of the cube, as such is at the state of zero value hyper cube 4.
12. In this light, when the description of the idol of Lord Brahma, the four head Lord, the presiding deity of 4-space is chased further, it shall be leading us to the fact as that the phenomenon of meditation of Lord Brahma, in cavity of his heart, as seat of his Lord, Lord Shiv, five head Lord and the presiding deity of transcendental worlds (5-space) results into Lord Brahma multiplying as ten Brahmas.
13. This phenomenon is of the order of the emergence of transcendental worlds within cavity of heart of Lord Brahma and that Lord Brahma multiplies ten fold and envelops the transcendental worlds.
14. Parallel to this is, as that, the origin of 4-space is seat of 5-space and that 4-space plays the role of boundary of 5-space constituted by ten hyper cubes 4.
15. This chase, as such, leads us to the distinguishing feature of 3-space and 4-space as that 4-space plays the role of origin of 3-space while 5-space plays the role of origin of 4-space.
16. This distinguishing feature of 3-space and 4-space, as well coordinates simultaneous existence of 3-

space and 4-space but now as 4-space being in the role of origin of 3-space.

17. As per the previous distinguishing feature of 3-space and 4-space, as well, there has been a phenomenon of simultaneous existence of 3-space and 4-space but in that situation, the 3-space was in the role of boundary of 4-space.

18. As such it would be a blissful exercise to simultaneous view a pair of situations of simultaneous existence of 3-space and 4-space, firstly as that 3-space is boundary of 4-space and secondly as that 4-space is the origin of 3-space.

RECAPITULATION AND SUM UP

Distinguishing features of 3-space and 4-space

Sr. no.	3-space	4-space
First	3 dimensions	4 dimensions
Second	Linear dimensions	Spatial dimensions
Third	Spatial boundary	Solid boundary
Fourth	4-space as origin	5-space as origin

EXERCISES TO EXPRESS AND SHARE

11. Express the way the 4-space reality is coming within your comprehension in terms of the nature of origins of 3-space and 4-space.

12. Evaluate your comprehensions of 4-space by comparative distinguishing feature of the nature of the origins of 3-space and 4-space.

■

ASPECT-2.5
3-SPACE AS DIMENSION AND 4-SPACE AS DIMENSION

1. It may be recapitulated as that the dimensions of 3-space are of linear order, that is, 1-space is playing the role of dimension of 3-space.
2. Also it may be recapitulated as that the dimensions of 4-space are of spatial order, that is, 2-space is playing the role of dimension of 4-space.
3. Firstly the above coordination of '3-space and 1-space' as well as of '4-space and 2-space' as domains and dimensions shall be leading us to the formulation as that (N-2) space is to play the role of dimension of N space, for N=1, 2.
4. This coordination of pair of spaces (N-2) space and N space, holds for their generality even for the values of N = 1, 2, 3, 4, 5, 6 and so on.
5. The scriptures preserve this comprehension at more than one place. Lord Shiv, five head Lord, accepts three eyes in each of his five heads. Incarnations of Lord Vishnu, illustratively, Lord Scanda, six head Lord, accepts four eyes in each of its six heads.

6. The parallel phenomenon of formats of idols of these Lords is of the formats of hyper cube 5 and hyper cube 6 as set ups of solid order and hyper solid order respectively.
7. The format of hyper cube 5 / 5-space / transcendental worlds / Shiv Lok / space presided by Lord Shiv, five head Lord with three eyes in each head is of set up of 3-space in the role of dimension of 5-space.
8. The format of hyper cube 5 / 6-space / Vishnu lok / space presided by Lord Vishnu, accepting incarnation as six heads Lords with four eyes in each head is a set up of 4-space in the role of dimension of 6-space.
9. With this, another distinguishing feature of 3-space and 4-space emerges in the nature of the roles of 3-space and 4-space as dimension folds, with 3-space as dimension structuring 5-space and 4-space as dimension structuring 6-space.
10. The parallel phenomenon is of the external angles of polygons of total sum of two pi (four right angles). The internal angles sum up (N-2) Pi (angles). This ratio for all values of N as N Pi takes to (N-2) Pi. As a result stand coordinated N and (N-2) like (N space) and (N-2) space.
11. It may be recapitulated that the general formulation of coordination of N space and (N-2) space has its first particular case as coordination of 1-space and 3-space with 1-space as dimension of 3-space. The second particular case of coordination of spaces is 2-space and 4-space with 2-space as dimension of 4-space. It is in this

sequence, 3-space stands coordinated with 5-space and 4-space stands coordinated with 6-space.

12. This coordination of 3-space with 5-space is parallel to the set up of pentagon with its total angles sum being 5 pi and sum of internal angles being 3 pi.
13. Likewise coordination of 4-space with 6-space is parallel to the set up of hexagon with its total angles sum being 6 pi and sum of internal angles being 4 pi.
14. This, this way, leads to another distinguishing feature for distinguishing 3-space from 4-space as that 3-space as dimension fold structures 5-space while 4-space as dimension fold structures 6-space.
15. Within 2-space / plane, the printout for 5-space bodies is of the layout of pentagon, while the printout for 6-space body is of the layout of hexagons.

RECAPITULATION AND SUM UP

Distinguishing features of 3-space and 4-space

Sr. no.	3-space	4-space
First	3 dimensions	4 dimensions
Second	Linear dimensions	Spatial dimensions
Third	Spatial boundary	Solid boundary
Fourth	4-space as origin	5-space as origin
Fifth	3-space is dimension of 5-space	4-space is dimension of 6-space

EXERCISES TO EXPRESS AND SHARE

13. Express the way the 4-space reality is coming within your comprehension in terms of the nature of 3-space and 4-space as dimensions.
14. Evaluate your comprehensions of 4-space by comparative distinguishing feature of the nature of 3-space and 4-space as dimensions.

■

Chase organisation formats
to attain transition with continuity

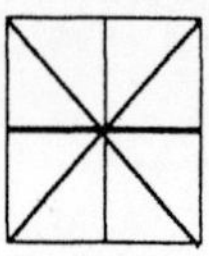

8 Symmetries of Square

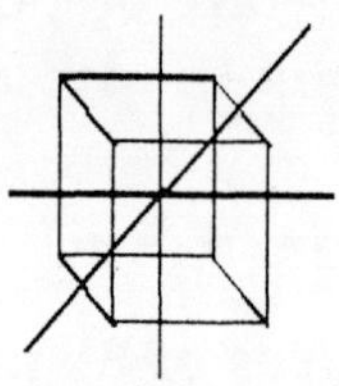

8 Octants cut of Cube

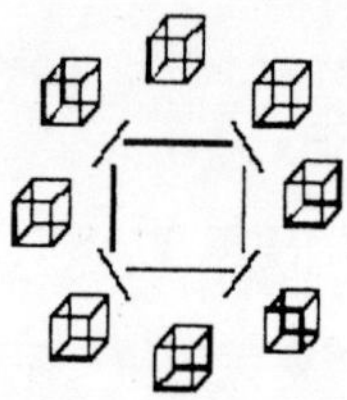

8 solid boundary components of hypercube-4

■

ASPECT-2.6
HYPER CUBES OF SOLID AND HYPER SOLID BOUNDARIES

1. It is the solid boundary of hyper cube 4, which makes it inaccessible.
2. It is this inaccessibility of 4-space because of solid boundary, which rebounds back to the permanence of mental blocks because of the physical world.
3. The par excellence feature of Vedic systems lies in its success of technique of transcending mind transcending through the solid boundary of Creator's space and making it lively as an intelligence field.
4. The scriptures preserve this feature and value of the transcendence technique of Vedic systems. The enlightenment of Idol of Lord Brahma, Creator the Supreme, takes us a step ahead as to how the transcending mind takes a step ahead of Creator's space (4-space) to transcendental worlds of Lord Shiv (5-space).
5. The transcendence technique takes transcending mind through solid boundary of Creator's space to Creator's space itself and further takes the transcending mind though hyper solid boundary

of transcendental worlds to the transcendental worlds themselves.

6. The phenomenon of transcendence through solid and hyper solid boundaries is the feature, which also is of different phases and stages.
7. The phase and stage of transcendence through solid boundary of Creator's space and reaching Creator's space itself is the phenomenon of creation of intelligence field by the transcending mind while transcending through the solid boundary of Creator's space.
8. The phase and stage of transcendence through the hyper solid boundary of transcendental worlds and reaching transcendental world themselves is the phenomenon of firstly attaining perfection of intelligence for the transcending mind and secondly creation of consciousness field for the transcending mind for the stage of reaching the transcendental worlds.
9. The phase and stage of transcendence through hyper solid boundary role being played by the Creator's space itself for transcendental worlds is the phase and stage of emergence of consciousness field within intelligence field itself.
10. As such this phenomenon of transcendence through solid and hyper solid boundaries deserves to be chased as a continuous phenomenon but of different phases and stages, with its first phase and stage being that of creation of intelligence field with transcendence through solid boundary and reaching the Creator's space. The second phase and stage is that of creation of consciousness

field as emerging within the intelligence field itself but at the phase and stage of perfection of intelligence.

11. Therefore this range of transcendence and consequential emergence of intelligence and consciousness field deserves to chase firstly as transcendence through solid boundary and reaching the Creator's space and having creation of intelligence field within Creator's space. Then the continuous glimpsing of the features of the Creator's space, the transcending mind sequentially attains perfection of intelligence. With it a phase and stage arises for further take off of the transcending mind for further transcendence through the hyper solid boundary and reaching the transcendental worlds. With this take off and transcendence, there emerges a consciousness field, which itself is of different states leading up till the unity state of consciousness.
12. As such the distinguishing features of 3-space and 4-space as in the roles of solid and hyper solid boundaries respectively of Creator's space and transcendental worlds is that of intelligence field and consciousness field with index of the intelligence field within Creator's space being of the order of perfection of intelligence on the other hand the index of consciousness field within transcendental worlds being of the order of unity state of consciousness.
13. It is this distinguishing feature of intelligence field of index of perfection of intelligence and of consciousness field of index of unity state of

consciousness which deserves to be chased time and again as it not only covers the solid and hyper solid boundaries but also takes through the Creator's space and the transcendental worlds.

14. The inter connection and coordination of 3-space and 4-space as solid and hyper solid boundaries of Creator's space and transcendental worlds, on the one hand, and as 'intelligence field of index of perfection of intelligence' and as 'consciousness field of index of unity state of consciousness', which ultimately becomes the dominating feature in terms of which the mundane worlds and the transcendental worlds are to be distinguished within Creator's space.
15. The creation of intelligence field and consciousness field by the transcending mind is the unique feature of Reality of the set up of the pursha format of human existence and for its knowledge and comprehension, everyone is to fully try and achieve.

RECAPITULATION AND SUM UP

Distinguishing features of 3-space and 4-space

Sr. no.	3-space	4-space
First	3 dimensions	4 dimensions
Second	Linear dimensions	Spatial dimensions
Third	Spatial boundary	Solid boundary
Fourth	4-space as origin	5-space as origin
Fifth	3-space is dimension of 5-space	4-space is dimension of 6-space
Sixth	Transcendence through solid boundary creates Intelligence field	Transcendence through hyper solid boundary creates consciousness field

EXERCISES TO EXPRESS AND SHARE

15. Express the way the 4-space reality is coming within your comprehension in terms of the nature of fields created while transcendence through solid and hyper solid boundaries.
16. Evaluate your comprehensions of 4-space by comparative distinguishing feature of the nature of fields created while transcendence through solid and hyper solid boundaries.

■

Chase the domains wrapped within boundaries

It would be intellectually satisfying to chase solid and hyper solid domain of cube and hyper cubes 4, 5 and 6 as wrapped wihtin spatial, solid and hyper solid boundaries parallel to the domain-boundary formulation

$$A^N : 2N\ B^{N-1},\ N=1, 2, 3, 4.$$

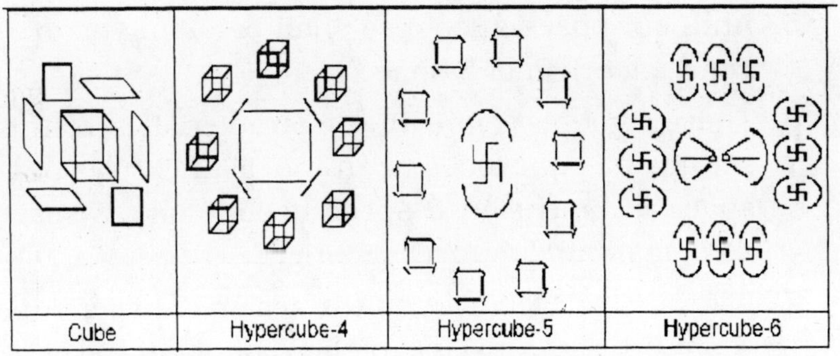

■

ASPECT-2.7
3-SPACE AND 4-SPACE IN THE ROLES OF ORIGIN FOLDS

1. Let us observe the set up of cube again.
2. Our observation may take us to 1-space playing the role of dimension for this set up.
3. Further our observation may take us to 2-space playing the role of boundary of this set up.
4. Still further our observation is to take us to 3-space as domain fold responsible for volume of this set up of cube.
5. And, our observation also shall be taking us to 4-space as origin of 3-space.
6. These four distinguishing characteristics of this set up of a cube focus upon the distinct roles of 1-space as dimension, 2-space as boundary, 3-space as domain and 4-space as origin.
7. Likewise, our focus upon the set up of hyper cube 4 shall be leading us to the distinct roles of 2-space as dimension, 3-space as boundary, 4-space as domain and 5-space as origin of 4-space.
8. The comparative study of these four folds set ups

for cube and hyper cube 4 shall be leading us to the following features:

Space	Dimension	Boundary	Domain	Origin
3-space	1-space	2-space	3-space	4-space
4-space	2-space	3-space	4-space	5-space

9. In the light of above sequential progression features, we may reach at a general formulation for N space as follows, which for N=3 and 4 shall be of the above features of 3-space and 4-space set ups.

Space	Dimension	Boundary	Domain	Origin
N-space	(N-2) space	(N-1) space	(N) space	(N+1) space
3-space	1-space	2-space	3-space	4-space
4-space	2-space	3-space	4-space	5-space

10. Now, as we are interested in the roles of 3-space and 4-space as origins, as such we can put the value for N+1=3 and N+1=4 respectively, and we shall be reaching at the following tabulation for 3-space and 4-space as origins.

Space	Dimension	Boundary	Domain	Origin
N-space	(N-2) space	(N-1) space	(N) space	(N+1) space
2-space	0-space	1-space	2-space	3-space
3-space	1-space	2-space	3-space	4-space

11. This distinguishing feature, is also leading us to another prominent feature of Vedic comprehension of whole range of dimensional spaces being compactified at same origins seat.

12. The sadkhas fulfilled with intensity of urge to know and to fully chase this comprehension of Vedic

systems, shall permit their transcending mind to glimpse this phenomenon of 4-space being the origin of 3-space, 5-space being the origin of 4-space and so on.

13. And as the dimensional orders of dimensional spaces being in sequential progression as much as that 1-space plays the role of dimension of 3-space, 2-space plays the role of dimension of 4-space, 3-space plays the role of dimension of 5-space, 4-space plays the role of dimension of 6-space and so on, as such there is inward expansion.

RECAPITULATION AND SUM UP

Distinguishing features of 3-space and 4-space

Sr. no.	**3-space**	**4-space**
First	3 dimensions	4 dimensions
Second	Linear dimensions	Spatial dimensions
Third	Spatial boundary	Solid boundary
Fourth	4-space as origin	5-space as origin
Fifth	3-space is dimension of 5-space	4-space is dimension of 6-space
Sixth	Transcendence through solid boundary creates Intelligence field	Transcendence through hyper solid boundary creates consciousness field
Seventh	Glimpse the phenomenon of 4-space as origin of 3-space, 5-space as origin of 4-space and so on.	

EXERCISES TO EXPRESS AND SHARE

17. Express the way the 4-space reality is coming

within your comprehension in terms of the nature of compactification of dimensional spaces at same origin seat.

18. Evaluate your comprehensions of 4-space by comparative distinguishing feature of the nature of compactification of dimensional spaces at same origin seat.

■

Chase center of the cube as ninth seat for 3 dimensinal frame

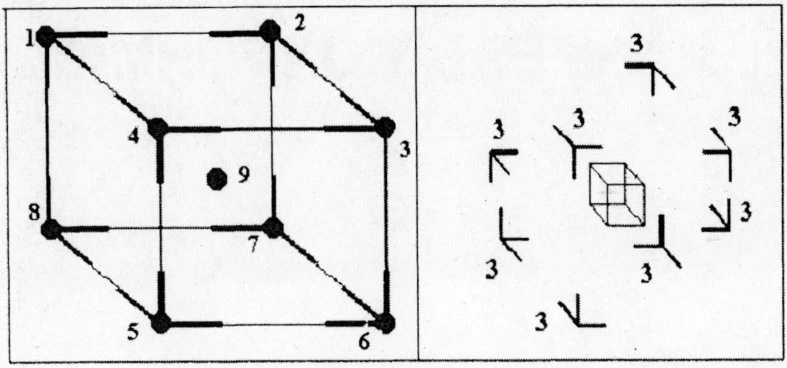

■

FEATURE-3
MANIFESTATION FORMAT

LINKING NOTE

FIRST PAIR OF FORTNIGHTS

Features

First fortnight

1. Feature 01 '4-space Reality'

2. Feature 02 'Distinguishing features of hypercube 4'

Second fortnight

3. Feature 03 'Manifestation formats'

4. Feature 04 'Transcendental base'

INTRODUCTORY

FIRST PAIR OF FORTNIGHTS

First pair of fortnights features focus upon 'FOUR SPACE REALITY' with concentration upon hyper cube 4, and of it, the feature of utmost interest being its origin fold / transcendental core of origin of 4-space as a seat of compactification of whole range of origins.

This takes us to the Sathapatya foundation. Its symbolic expression may be as under, a set up of Shivligam / transcendental stem / pole installed within Creator's space at its transcendental core.

3RD FEATURE

The **third feature** is about the manifestation formats of the entities around us. **Fourth feature** is going to be concerning the origin folds compactification at origin of four space as seat of five space.

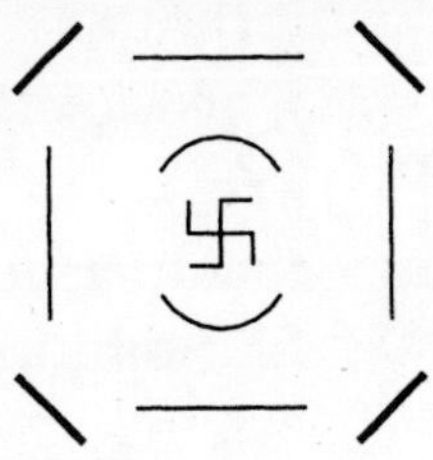

ASPECT-3.1
FOUR SPATIAL DIMENSIONS

INTRODUCTORY

1. We are entering from today, the second fortnight of the course.
2. Two features covers so far are four space realty and the distinguishing features of hyper cube 4 over cube.
3. We have tried five distinct approaches to reach 4-space. Also we have been through seven distinguishing features of cube and hyper cube 4.
4. Here, during their fortnight, we are aiming to take another pair of features, namely, manifestation format and the 'transcendental base' of these formats.
5. It would be great, if pair of fortnights' features, are viewed simultaneously as four features together going to give us an insight of Sathapatya foundation as of 'hypercube 4 with transcendental origin'.

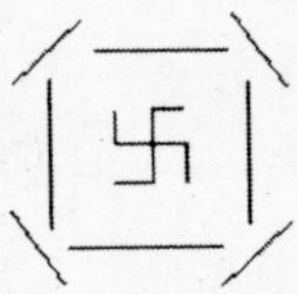	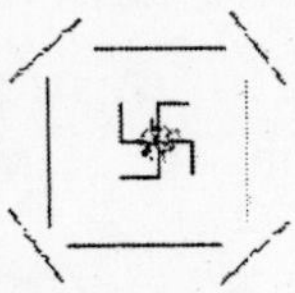		
Spatial order	Transcendental origin	Five space set up within four space	Solid order

6. This phase and stage of the course takes us to the stage of chase of 4-space reality along manifestation formats of Creator's space, beginning with 4-space spatial dimensions'.
7. The 4-space reality prominently distinguishes itself from 3-space, by the nature of its spatial dimensional order.
8. The dimensional frame of 4-space avails a set of four spatial dimensions while three-dimensional frame is a set up of just three linear dimensions.
9. Before chase step is taken, it may be relevant to have a fresh look at representative symbolic expression of hyper cube 4, viz

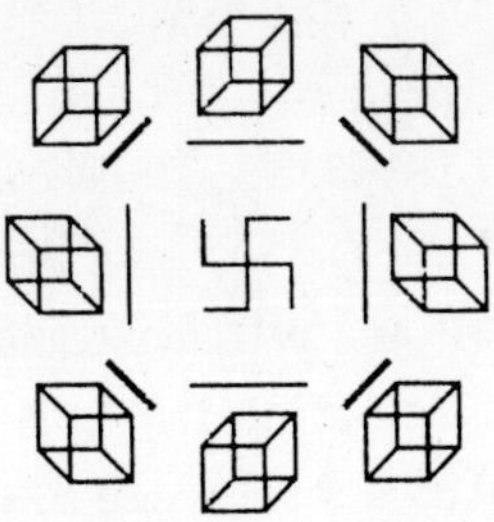

FOUR SPATIAL DIMENSIONS

10. The focus here is upon the dimensional frame of

four space as of four spatial dimensions.

11. This also may be viewed as a measuring rod of four steps as a synthetic set-up of 'interval, square, cube and hypercube 4'.

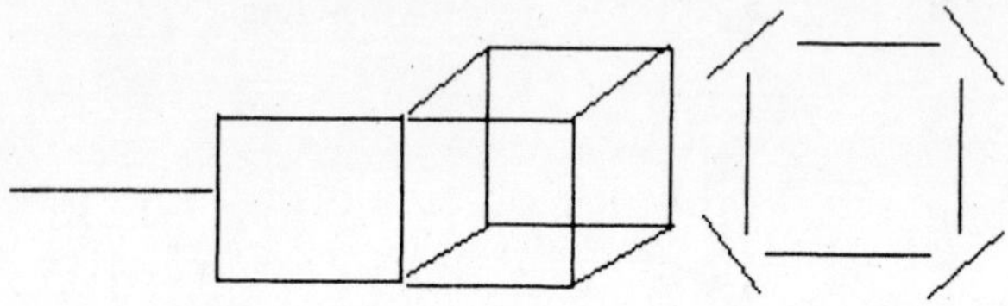

12. This, as much, is to be as 'a manifested path' of four steps, where at every another step, a new set-up is available.
13. This four fold path is of 'manifestation format' with a track for the 'first step being that of a 'monad', the second step being a track of a 'di-monad', third step being a track of a 'tri-monad' and fourth step being a track of a tetra-monad' formally may be defined as a 'point'; without parts. Di monad may be taken as a point of two parts; and tri-monad and tetra monad being respectively of three and four parts.
14. As such, each part is a track of a 'monad' of with or without parts.
15. When, 'monad' is 'without parts' it is taken as leading to linear order paths, while it is of parts, it is of synthetic set ups for its tracks.
16. The di-monad, being of 'two parts' point, its track is taken as of spatial set-up.
17. The availability of this (spatial) order within 4-

space, shall be making it importance to accept according the transition and transformation into the set-ups of the tracks at all the four steps of manifested path.

18. It is just like shifting from 1 as 1 to 2 as 1.
19. It shall be making 'monad' as of a pair of parts.
20. So '1' may be taken as a pair; 2 as double pair; 3 as triple pair and 4 as quadruple pair.
21. This, in a way, would amount to be the set-ups of features, at first step, as of one spatial dimension and of subsequent, second, third and fourths, as to be of double, triple and quadruple pair' respectively.
22. This may be viewed, as dimension of two parts, at every step.
23. This also may be viewed as that dimension is a pair of half dimension; dimensional frame as pairs of dimensional frames of half dimension.
24. And, accordingly, the 'origin' of 'dimensional frame' as well becomes of features sustaining pair of dimensional frames of half dimensions.
25. This transition and transformation at level of dimension leading to transition and transformation at level of origin as well, is a 'feature' which deserves to be chased well.
26. This chase can be 1=1/2+1/2 and 0= 0+0.
27. ½ +½=1; (½, ½); (A, A); (A,A) =2A; 2.
28. With it

    1=2

    2=1+1=2

3=(1½, 1½)=2

4=(2, 2)=2

5=(2½, 2½)=2

29. The manifested path to emerge.

1	2	3	4
		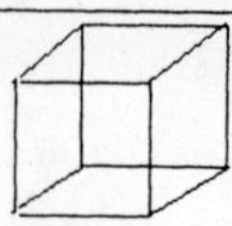	
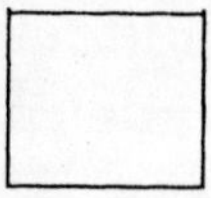	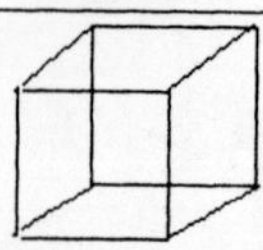	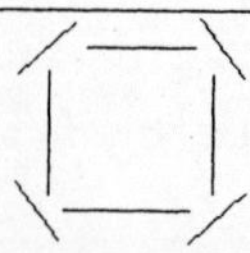	
Square	Cube	Hyper cube 4	Hyper cube 5
2-space	3-space	4-space	5-space

■

Further blissful exercise

To reach at the the next step of the manifestation format withcube / 3-space as dimension, hyper cube-4 / 4-space as boundary, transcendental worlds / 5-space / hyper cube-5 as domain and the self referral domain / 6-space / hyper cube-6 as origin base.

■

ASPECT-3.2
FOUR STEPS LONG MEASURING ROD

1. 'Four spatial dimensions' Set-up, that is, four space domain, accepts a four steps long measuring rod constituted by 'interval, square, cube and hypercube 4'.
2. This 'measuring rod' helps dive deep into the 'four space domain'. The 'measuring rod', as such emerges a 'diving ladder' and the four steps of the measuring rod being the four steps of the diving ladder'.
3. The spatial dimensional order of '4-space' is acceptance of '2' as '1'; 'point' as 'di-monad', 'pair of axes' as axis' and each step of diving ladder being of four quarters Set-up parallel to four steps of a measuring rod accepting full structure of '4-space domain' within its 'every point'.
4. These features of the 'Set-up of the 4-space domain' may be chased by 'diving mind', as availing 'pairing' of points / ions / monads / elements, along 'measuring rod format' as Ganita Sutras / mathematics of dimensional folds for artifices of elements, 26 in number of range 1 to 26, being

meters of four quarter each, and also, together these to be an alphabet of 26 letters, pairing alphabet letters of 'alphabet of 52 letters' and further pairing the paired letters and having full-fledged coverage for whole range of all the 104 quarters of all the 26 meters.

5. This, as such, bring within the expected range of learning to dive deep along diving ladders steps availing measuring rod (1), learning pairing of artifices '1 to 26' for the alphabet 'A to Z' as range of artifices '1 to 26' with 'A' as '1' and 'Z' as '26'. And others B to Y being 2 to 25 in that sequence and order (2) and for their mathematics, science and technology learning Ganita Sutras (3).
6. Pairing of artifices '1 to 26' shall be giving us an insight as to how the words-formulations of orthodox and classical 'English' are composed and formulated to chase the Set-up of the Creator's space, and the same (these words formulations) is going to be our 'big treasure' find, worth the order of the 'steps of diving ladder' which shall be leading up till the CENTRE OF FOUR SPACE, where is located a seat of transcendental worlds.
7. Insight of sweetness of bliss of the pairing process of artifices 1 to 26 as 'basic elements' may be had by having chase of the words formulation by adding positions of letters along the range 1 to 26, with A=1, B=2,− −Z=26 as their sum value as number value format of the cone which 'word' being a formulation. Word 'ladder' is availing six letters, namely, (1) L, (2) A, (3) D, (4) D, (5) E, (6) R. These are of position values:

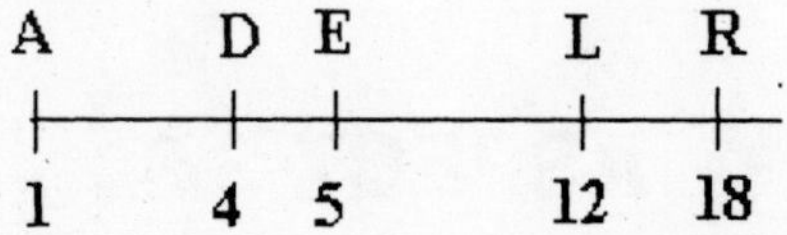

8. Number value format of these word for mutations, namely, LADDER comes to be, as sum of the position values of these letters respectively as 12+1+4+4+5+18 = 44, which is of the order of SPACE, as S =19, P=16, A=1, C=3, E=5 sum up 19+16+1+3+5 =44.
9. And STEP yields S=19, T=20, E=5, P=16 together 19+20+5+16=60. And it is of the order of 'FOUR' as F=6, O=15, U=21 & R=18 yields 6+15+21+18=60 so 'LADDER STEP' is '4-space'.
10. As such, 'ladder step' as '4-space' is of 'spatial order' and it is availing four quarters format'.
11. 'Interval' as 1-space body / domain is of four quarters (-1 space, 0 space, 1-space, 2-space), in short, (-1, 0, 1, 2) ending into 2-space / artifice 2. And 'square' as (0, 1, 2, 3) shall be ending into 3-space / artifice 3. Like was 'cube' as (1, 2, 3, 4) step shall be leading to 4-space and hypercube 4 shall be leading to (2, 3, 4, 5), that is 5-space.
12. Diving ladder takes up till 'center' of 'hyper cube 4' / transcendental core of 4-space / seat of transcendental worlds / consciousness field as NVF 'dive deep'=70 as STOP as NVF (STOP)=70. It may be expressed as that the 'STOP' of 'DIVE DEEP' along Diving ladder is the transcendental core of the Creator's space.

■

ASPECT-3.3
FOUR FOLD MANIFESTATION FORMAT

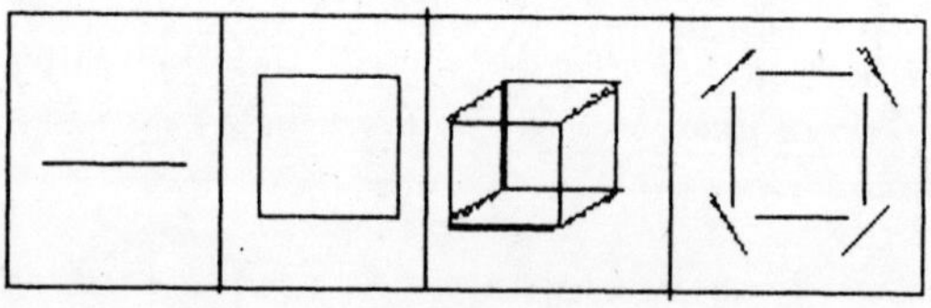

1. Interval (—) as first step of the 'measuring rod' / first quarter of the manifestation format / first step of diving ladder' with spatial order of 4-space makes it a set-up of a manifestation layer of four parts, parallel to four parts of pair of areas.

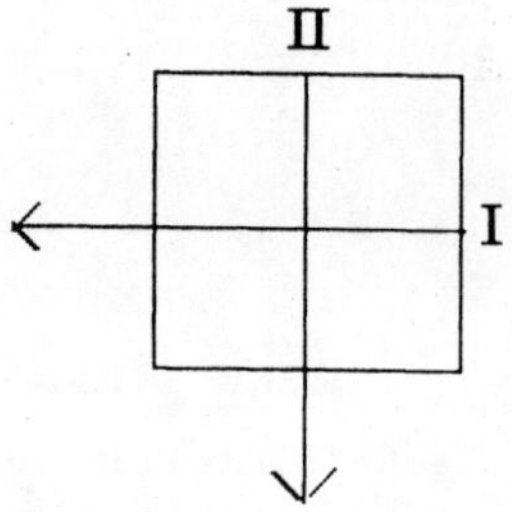

2. It uniquely remains linear and spatial both, being 1-space within 2-space.

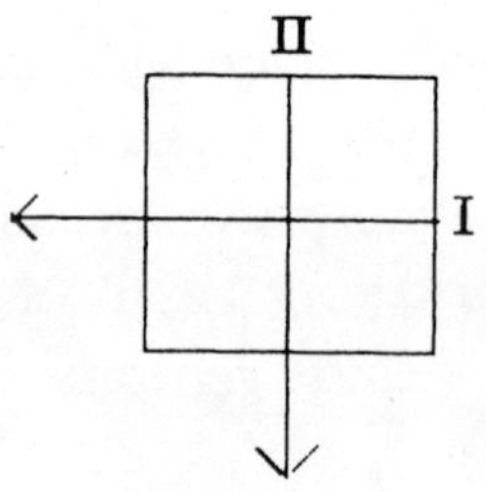

3. It ultimately takes a pair of flow forms through center / origin

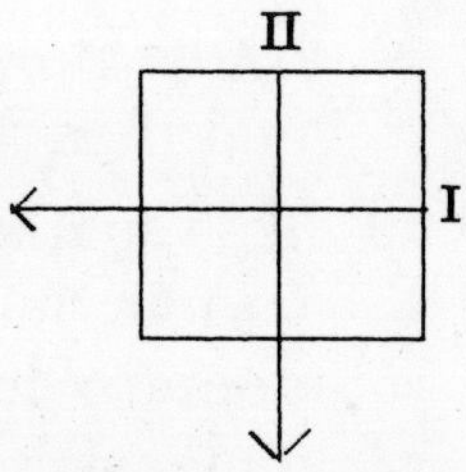

with either of axes, to be first, shall be supplying a pair of paired streams.

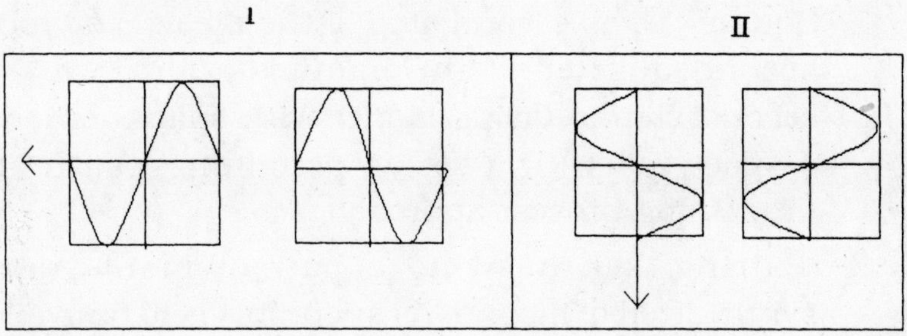

4. This pair of paired streams makes 'four' as of prime focus 'here'. The coverage of all the 'four planes' takes us to 'four space' as 'linear progression' carrying spatial order as 'wave', splitting as 'halves' and 'halves' splitting as 'quarters'.
5. This chase of spatial order as sequential progression is a diving path of the transcending path of mind. As such, first aspect of manifestation format, which need to be learnt is how to chase the diving path of 'transcending mind'. It is this exercise, which is to be at the level of mind itself, which is basically performed by the mind availing one axis for 'diving' and other for its self-referral chase. And in its process, both pairs of paired streams of both axes come into processing, and the self referral processing process emerges where

by the 'processing –process' firstly splits into halves and then 'halves' are split into 'quarters'.

6. This spatial order linear progress gets coordinated as four fold viz. (dimension fold, boundary fold, domain fold and origin fold) / (-1) space as dimension fold, 0 space as boundary fold, 1-space as domain fold and 2-space and 3-space / (-1, 0, 1, 2).
7. Square (■) as second step of the measuring rod / second quarter of the manifestation format / second step of diving ladder with spatial order of 4-space makes it a set up permitting 'pairing' as processing-process approach.
8. 'Pairing' synthesizing a pair of monad as a synthesized di monad. As such, this is to be availed as the phase and stage of learning pairing process, as processing process of spatial order of 4-space.
9. The catchword here is a 'fold'. How the order folds a monad being a synthesized set of 'halves' and folds a half of a monad as a pair of quarter. Linear progression within 4-space may be chased as a moving track of a 'solids' as 13th edge of 'cube'. The spatial order provides a spatial format for 13th edge. A pairing there shall be providing as artifice of 26, being 13+13.
10. This 'pairing' of 'paired one's' is also an intellectual skill whose proficiency as well would be blissful as it is going to be and whole range of features of Creator's space shall be unfolded and also the whole range of words-formulations of orthodox and classical English as well as unfolding, with NVF (ENGLISH) =74= NVF (PAIRING); pairing of folds,

as NVF (PAIRING)= NVF (FOLD)+ NVF (FOLD). And then, 'FOLD' is 'CONE', as NVF (FOLD) = NVF (CONE)=37.

11. MANIFESTATION

NVF (MANIFESTATION) =126=63+63

=NVF (LIMIT) +NVF (LIMIT)

And (63) is a pair of reflection artifices; 6=6 & 3=3, it is a Dev-nagri script for 6=6, and 3=3. This shall be taking us to a split up an axis of cube a split process of cube a split process of cube as a pair of cones / folds of opposite orientations.

12. ION BEAM VOID CONE

Let us chase MANIFESTATION as 'MANI FEST AT ION' with NVF'S sequence '37, 50, 21, 38'. This supplies of same generic formulations quadruple 'CONE, VOID, BEAM, ION) which with reversal of orientation (+1 space as domain leading to –1 space as domain leading to –1 space as dimension) as ION BEAM VOID HALF.

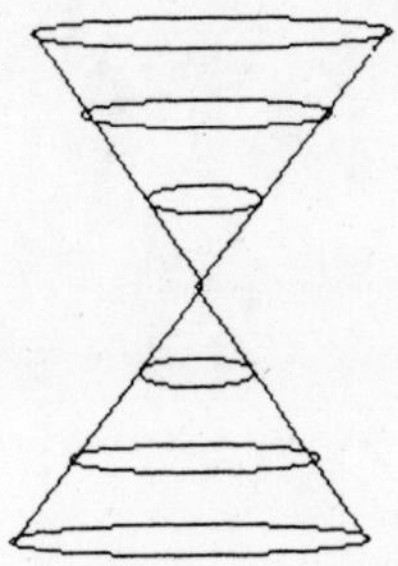

13. It is this feature of MANIFESTATION to see that only 'CONE' of paired 'CONES' to go 'void' deserves to be chased. It is this feature, which is at the base of the calculus availing only half-boundary; A^N: $2NA^{N-1}$ is a domain –boundary formulation, and NA^{N-1} is the first derivative.

14. This, as such, is the phase and stage of learning 'Sutras / formulas'; very first of which 'one more that its previous one' sets a process of ordering by

'pairing' (one, one), leading 'a' to A 'a', '1' as'0+1'; and also, a chase of 'a thing' as 'thing', that is, a 'sequential progression' as 'attain set up' with a difference of just '1', which shall be taking 126 to 125 or other way round '125 to 126'. And, 125 = 5 x 5 x5, a cubic superstructure upon a square format 5 x 5.

15. A sequential progression 5^3 and affine set up 5^2 are so inter connected that this coordination there of deserves to be comprehended by having concentrative look at:

1 1 1 1 1	1 2 3 4 5
1 1 1 1 1	2 3 4 5 6
1 1 1 1 1	3 4 5 6 7
1 1 1 1 1	4 5 6 7 8
1 1 1 1 1	5 6 7 8 9
= 5 x 5 =25	= 125 = 5 x 5 x 5

16. FORMAT

NVF (FORMAT) =73=NVF (CONE)= 37

NVF (CONE FORMAT)

= (37, 73, a reflection pair of artifices.

=110=55+55=(55, 55)

=NVF (SKY, SKY), a paring of SKY, which is to 'format cone' as 'sphere angle' / LAYER AXES.

17. This way FOUR FOLD MANIFESTATION FORMAT attains 'layer-axes' for manifestation layers / manifestation creator / (counting, counting) NVF (MANIFESTATION LAYER) = 206 = 126+ 80 = 103+103.

■

ASPECT-3.4
INTERVAL AS A MANIFESTATION LAYER (-1, 0, 1, 2)

EXERCISES-1

1. This and the following three days are to be availed for exercises and expressions for self-evaluation of one's comprehensions about the first fold of manifestation format as availing full manifestation format.
2. One view would be to approach first fold as interval as a manifestation layer (-1, 0, 1, 2).
3. This fold of manifestation format may also be viewed and chased as 'first step of 'diving ladder'.
4. It would be a blissful exercise to chase first quarter of measuring rod as
   (a) Interval,
   (b) Manifestation layer (-1, 0, 1, 2)
   (c) 'First step of diving ladder'
   (d) 'Linear progression'
   (e) 'Linear progression of spatial format'
   (f) 'Sequential progression of spatial format'
   (g) 'Sequential progression of solid format'

(h) 'Sequential progression of hyper solid format'

5. Further it would be a blissful exercise to permit the transcending mind to dive and at the end of scaling of first step along diving ladder to attain up till its end as transition and transformation for the second step scaling along the diving ladder.
6. One shall chase once comprehension of the diving attainments of the transcending mind and to reach at ones own evaluation of one's comprehension of one's transcending mind.
7. One's evaluation of one's comprehension may be shared with fellow sadkhas and seniors, as these evaluations are self certifications and when those stand shared with other ensure satisfaction of attainments.

■

Comparative chase

This phase and stage may be availed for comparative chase of the present day approach to the set up of a line / interval / 1-space in the light of the ancient wisdom of Vedic systems.

The distinguishing difference may be appreciated as that while the present systems restrict themselves only up till the first linear axis of 3-space being the whole set up of 1-space, while on the other hand the ancient wisdom approaches it as a manifestation layer and that (-1) space playing the role of dimension of (+1) space.

■

ASPECT-3.5
SQUARE AS OF SPATIAL FORMAT AND ALSO AS A MANIFESTATION LAYER
(0, 1, 2, 3)

EXERCISES-2

1. One view would be to approach second fold as square of spatial format and as a manifestation layer (0, 1, 2, 3).
2. This fold of manifestation format may also be viewed and chased as 'second step of 'diving ladder'.
3. It would be a blissful exercise to chase second quarter of measuring rod as
   (a) Square,
   (b) Manifestation layer (0, 1, 2, 3)
   (c) 'Second step of diving ladder'
   (d) As a process of pairing of artifices
   (e) As a process of pairing of monads
   (f) As a process of pairing paired monads / synthesized monad
   (g) As a processing process of folds
   (h) As a pairing for axes of solid order as a pair of folds / cones
4. Further it would be a blissful exercise to permit

the transcending mind to dive and at the end of scaling of second step along diving ladder to attain up till its end as transition and transformation for the third step scaling along the diving ladder.

5. One shall chase once comprehension of the diving attainments of the transcending mind and to reach at ones own evaluation of one's comprehension of one's transcending mind.
6. One's evaluation of one's comprehension may be shared with fellow sadkhas and seniors, as these evaluations are self certifications and when those stand shared with other ensure satisfaction of attainments.

■

Comparative chase

For proper insight about the set up of 2-space / surface / plane / square, one shall have its comparative chase along with interval and cube, along the following manifestation format.

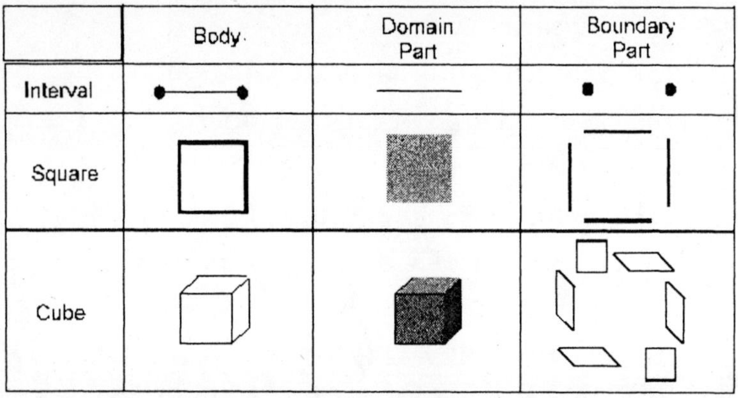

■

ASPECT-3.6
THIRD FOLD ALSO ACCEPTS FULL MANIFESTATION FORMAT

EXERCISES-3

CUBE AS OF SOLID FORMAT
AND ALSO AS A MANIFESTATION LAYER
(1, 2, 3, 4)

1. One view would be to approach third fold as cube of solid format and as a manifestation layer (1, 2, 3, 4).
2. This fold of manifestation format may also be viewed and chased as 'third step of 'diving ladder'.
3. It would be a blissful exercise to chase third quarter of measuring rod as
   (a) Cube,
   (b) Manifestation layer (1, 2, 3, 4)
   (c) 'Third step of diving ladder'
   (d) As a manifestation process
   (e) As a process of dominant 'domain fold'
   (f) As a process of 'pair of folds / cones' along each axis

(g) As 'ion beam void cone'

(h) As a solid sequential progression along an affine spatial format

(i) NVF (MANIFESTATION LAYERS) = NVF (COUNTING, COUNTING)

4. Further it would be a blissful exercise to permit the transcending mind to dive and at the end of scaling of third step along diving ladder to attain up till its end as transition and transformation for the fourth step scaling along the diving ladder.
5. One shall chase once comprehension of the diving attainments of the transcending mind and to reach at ones own evaluation of one's comprehension of one's transcending mind.
6. One's evaluation of one's comprehension may be shared with fellow sadkhas and seniors, as these evaluations are self certifications and when those stand shared with other ensure satisfaction of attainments.

■

Evaluate your comprehension of manifestation formats of solids

It would be a blissful exercise to chase solids for their common manifestation format.

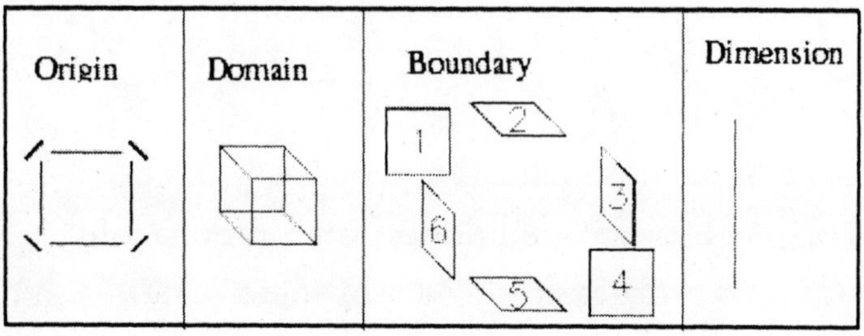

■

ASPECT-3.7
FOURTH FOLD ALSO ACCEPTS FULL MANIFESTATION FORMAT

EXERCISES-4

HYPERCUBE 4 AS OF HYPER SOLID FORMAT AND ALSO AS A MANIFESTATION LAYER (2, 3, 4, 5)

1. One view would be to approach fourth fold as hypercube 4 of hyper solid 4 format and as a manifestation layer (2, 3, 4, 5).
2. This fold of manifestation format may also be viewed and chased as 'fourth step of 'diving ladder'.
3. It would be a blissful exercise to chase fourth quarter of measuring rod as
   - i. Hypercube 4,
   - ii. Manifestation layer (2, 3, 4, 5)
   - iii. 'Fourth step of diving ladder'
   - iv. As a transcendence process
   - v. As a process of dominant 'origin fold'
   - vi. As a process of 'reaching transcendental core'
   - vii. As a process of diving deep and glimpsing the 'transcendental worlds'
   - viii. As a phenomenon of compactification of origins

of whole range of dimensional spaces at transcendental core

ix. As a phenomenon of spatial order encircling the transcendental core

x. As a phenomenon of emergence of solid quantifiers within hypercube 4 through its transcendental core

xi. As a phenomenon of manifestation of solid quantifiers within 4-space

xii. As a phenomenon of emergence of 5-space as transcendental universe within Creator's space

xiii. As a phenomenon of split of transcendental universe as a pair of distinct transcendental worlds around transcendental sky within Creator's space

xiv. As a phenomenon of sustenance of transcendental worlds within Creator's space by 'self referral domains' of 6-space as origin source reservoir of transcendental worlds (5-space)

4. Further it would be a blissful exercise to permit the transcending mind to dive and at the end of scaling of fourth step along diving ladder to attain up till its end as transition and transformation leading to the self referral domains of 6-space as origin source reservoir of transcendental worlds compactified at transcendental core of Creator's space.

5. One shall chase once comprehension of the diving attainments of the transcending mind and to reach

at ones own evaluation of one's comprehension of one's transcending mind.

6. One's evaluation of one's comprehension may be shared with fellow sadkhas and seniors, as these evaluations are self certifications and when those stand shared with other ensure satisfaction of attainments.

■

Glimpse seat of origin as manifested format

4-space as origin of 3-space / cube, itself accepts 5-space as origin along the manifestation format. One shall glimpse this phenomenon. The symbolic expressions may be availed as under:

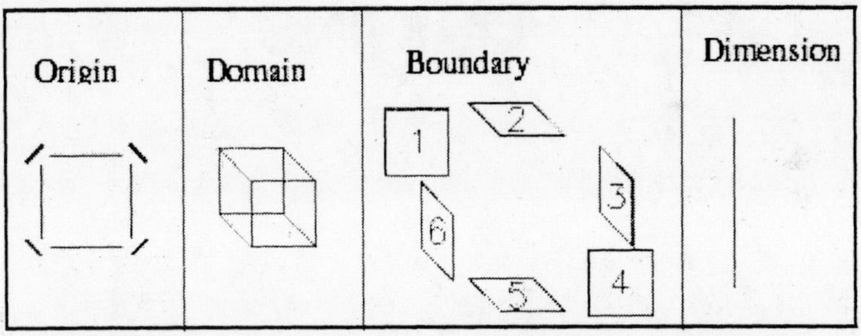

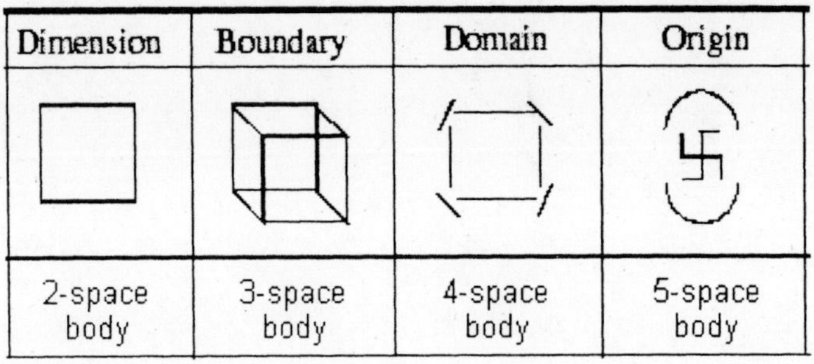

■

FEATURE-4
COMPACTIFICATION OF ORIGINS

FEATURE-4

Fourth feature is going to be concerning the origin folds compactification at origin of four space as seat of five space.

ASPECT-4.1
COMPACTIFICATION OF ORIGINS AT TRANSCENDENTAL CORE OF CREATOR'S SPACE

1. Chase here is about location and setting of 'origins' of the dimensional spaces.
2. Chase of 'origins' is as 'origin-folds' of the dimensional spaces as manifestation layers.
3. Origin-fold is the fourth fold of the manifestation layer, it is the fourth quarter of the manifestation format.
4. Each quarter of the manifestation format being a format for the manifestation of a dimensional space, therefore this chase is to be of dimensional spaces themselves but in the context of their specific roles as 'origin-folds' /origins.
5. This chase is in fact the chase in continuation of the chase of 'manifestation format'.
6. So, it as such, remains within Creator's space.
7. It takes us to its origin resource, transcendental core.
8. Therefore, first, we shall equip ourselves with presence conceptual comprehension of 'transcendental core of Creator's space'.

9. Measuring rod of this space shall be sequentially taking us to its fourth step, which is hyper cube 4 itself, the representative regular body of 4-space.
10. It is 'center' of hyper cube 4 which is availing the origin-seat of 4-space as the transcendental core for manifested entities within Creator's space.
11. As such, 'transcendental core' is the 'origin' of 4-space / origin seat for the transcendental worlds /5-space.
12. Therefore, it is a unique carrying of the di monad order through all the four steps of the measuring rod within Creator's space.
13. The 'spatial order' of the Creator's space, to begin with, as first step of the measuring rod, shall be unfolding at subsequent steps as solid order, hyper solid 4 order and finally being the hyper solid 5 order.

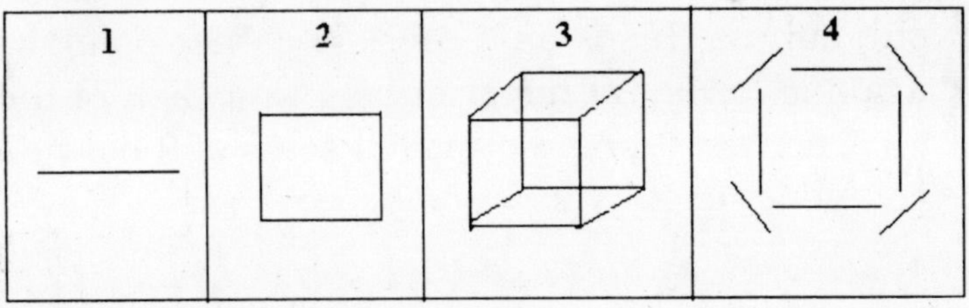

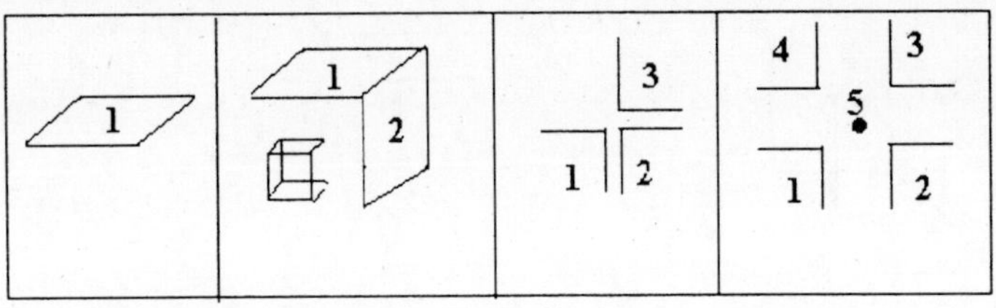

14. It is this 'transcendental order / 5-space in the roll of dimension as origin-fold within Creator's space' which becomes chase range of the

phenomenon of compactification of the origins of dimensional spaces.

15. Chase may be, first had, in reference to 3-space, in the context of 'cube', as a four fold manifestation layer (1, 2, 3, 4) as a linear dimensional order, spatial boundary, solid domain and hyper solid (4) origin.

16. 'Centre' of cube emerges as 'hyper solid 4' point / origin seat.

17. 'Centre' of 'hyper solid-4' as well would be 'seated' at this very location, and thereby, there being a compactification of centers / origin, as one superimposed upon other, as base there of.

18. Insight of this 'seat' may be had, firstly concentrating upon 'centre' of a 'cube' as the 'synthesise' point of 'eight sub cubes', as common location of one corner point of each of these eight sub cubes. This is as if these '8 corner points' are availing format of hyper cube 4 and each of these 8 corner point being provided seats of 8 boundary components of hyper cube 4.

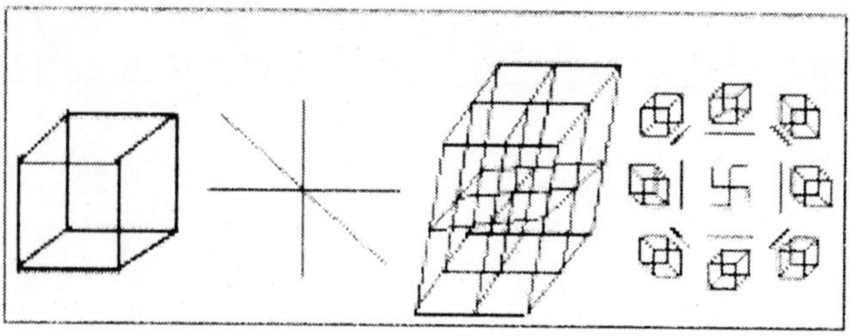

19. This shall be providing us insight of center of cube and hyper cube 4 being at same point and also that the centers of hyper cube 4 and hyper cube 5 as well are at the same very point.

ASPECT-4.2
SPATIAL ORDER LEADS TO COILING AROUND TRANSCENDENTAL CORE

COILING AROUND THE TRANSCENDENTAL CORE

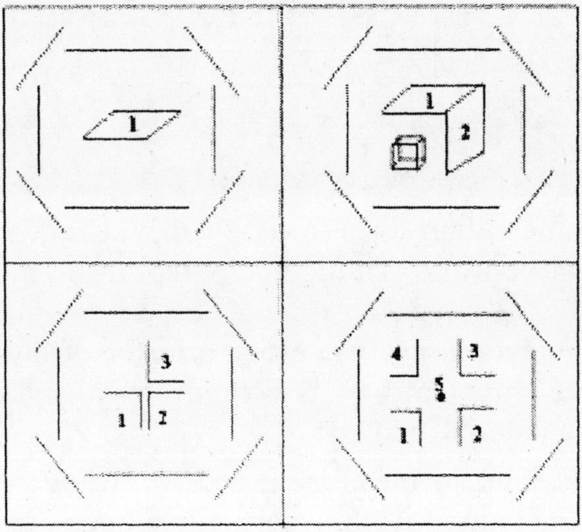

1. Insight gained as to the 'centers' getting setting at the 'center' of hyper cube-4, shall be helping us to glimpse further as that the 'spatial order leads

to coiling around the transcendental core of Creator's space'.

2. As such, this phase and stage of the chase of the phenomenon of compactification of origins at transcendental core of Creator's space, is to be focused upon as to how, it is the 'spatial order of the Creator's space which is leading to coiling around the transcendental core.
3. This phenomenon of coiling around the transcendental core is to be chased availing the measuring rod which shall be taking us to the origin fold.

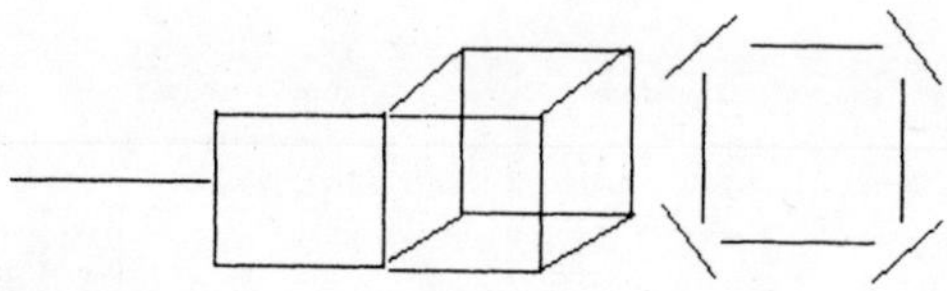

4. Manifestation format supplied by Creator's space shall be transforming each step of the measuring rod / as steps of 'diving ladder' as quarters of manifestation layer; every quarter of layer being a distinct fold of the layer availing distinct dimensional content.
5. First step of measuring rod works out four fold diving ladder availing contents of four consecutive dimensional spaces, namely (-1, 0, 1, 2) / (-1 space, 0 space, 1-space and 2-space).
6. Like that the second, third and fourth step, lead to manifestation layers of four folds each availing

the dimensional contents of four spaces at a time, viz., (0, 1, 2, 3); (1, 2, 3, 4) and (2, 3, 4, 5).

7. The domain fold corresponding to the third step of the measuring rod, here shall be a manifestation layer (1, 2, 3. 4); and 3-space content for its third quarter shall be making this fold of a split order of 8 components for the 'boundary' cover of the fourth quarter of the layer.
8. It is this feature of 'third quarter' of 'domain-fold' providing boundary cover for the fourth quarter thereof, which is to become responsible for the coiling around the transcendental core within Creator's space.

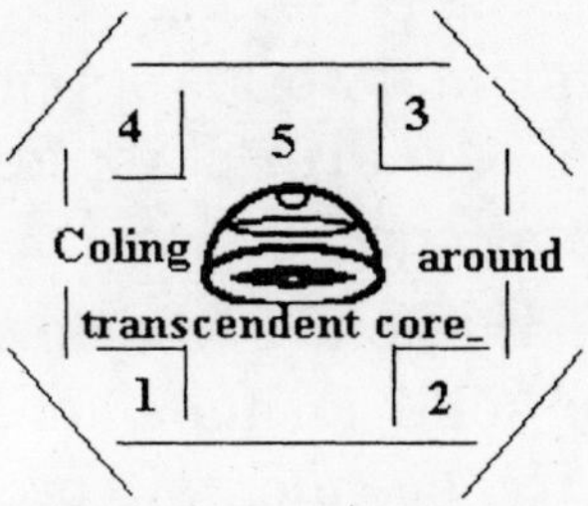

■

ASPECT-4.3
SOLID ORDER OF TRANSCENDENTAL CORE SUPPLY SPIRAL FORMATS FOR SPATIAL COILS

1. The coils around the transcendental core are of spatial format.
2. The transcendental core is the seat of solid order transcendental worlds.
3. The transcendental worlds supply solid quantifiers within the Creator's space through its transcendental core.
4. The solid quantifiers manifest as a stem / pole and provide solid formats for the spatial coils around the transcendental core.
5. The spatial coils because of its pair of axes flow streams have progressions as of parallel streams and also of a unified stream.
6. The parallel streams may be of identical orientations, as well as those may be of opposite orientations.
7. The unified stream as well may be of either of the four directional streams (north, east, south, west), and further those may be of four angular streams (north-east, east-south, south-west, west-north).

8. The solid format of stem / pole of manifested solid quantifiers, shall be, as a manifested solid waves, shall be of four quarters and of these, the second quarter as second fold shall be covering the stem / pole as a boundary fold and as such also would be providing a spiral format for the transcendental core coils.
9. The spatial boundary fold coverage of the stem / pole as a format for the spatial coils shall be working out a set up of the flow paths for the compactified origins as synthetic monads permitting pairing within Creator's space.
10. It is this feature of compactification of origins as synthetic monads permitting pairing within Creator's space, which need be chased as spiral formats for the spatial coils around the transcendental core of the Creator's space.
11. The sadkhas fulfilled with intensity of urge to fully know and to completely chase the phenomenon of origins as synthetic monads permitting pairing within Creator's space shall permit their mind to transcend time and again and to glimpse this phenomenon from the stage of splitting of the spirals around the transcendental stem with more focus and attention as it is to perfect the intelligence and also to attain unity state of consciousness for chase of emergence of 5-space Reality within 4-space.

■

ASPECT-4.4
SPATIAL ORDER ACCEPTING SPLITS FOR THE SPIRALS AROUND THE TRANSCENDENTAL STEM

SPLIT OF SPRIALS

1. The chase of 'split of spirals' needs special focus and attention for the reason as that it is a unique happening at whose base, both spatial order of the Creator's space and solid order of the transcendental worlds are simultaneously at work.
2. The manifestation of solid quantifiers as stem / pole at seat of transcendental core within Creator's space and there being a spatial cover for the stem because of the spatial order of the Creator's space, it amounts to having spatial superimposition upon the spatial boundary fold of the manifestation format of solid quantifiers.
3. It is this happening of spatial super imposition upon the spatial boundary fold of the manifestation format of solid quantifiers, which provides a distinct features for the solid quantifiers stem, as those of spatial grids superimposed upon each other.
4. The features of spatial boundary fold of solid quantifiers stem as of spatial grids super imposed

upon spatial grids, are of such characteristics which makes the existence phenomenon as of compactified states.

5. This phenomenon of existence as compactified states deserves to be chased fully as in terms of it only, the insight may be had into the reality of existence at different states within the same frame, like existence within human frame.
6. Existence within human frame as a compactified set of states may be chased in the background of insight into the emergence of super imposed spatial grids at the boundary of solid quantifiers stem.
7. This feature of acquiring super imposed spatial grid at the boundary, would make the boundaries of hyper solids as of superimposed solid grids; and then the superimposed solid grids for hyper solids shall be constituting a sequential progression range of solid and hyper solid manifested waves layers whose chase is to be of many fold bliss for the transcending mind chasing phenomenon of origins as synthetic monads.
8. This phenomenon of sequential progression range of grids super imposed upon grids splits into many limbs because of the availability of additional axes at each progression step.
9. This, as such deserves to be chased beginning with the split of spirals at the boundary of the solid quantifiers stem within Creator's space.
10. This split of spirals at boundary of solid quantifiers stem deserves to be chased axes wise, along each of the axes of the spiral, taking one axes being constantly supplied by the stem itself, and other

axes for the coils splitting spirals is to be supplied by the axes of the spatial order of the boundary of the stem.

11. The availability of super imposed spatial grids at the boundary of the stem, deserves to be chased grid by grid, and as a spatial axis by a spatial axis.
12. This as such shall be making this chase as of sequential features for the axes themselves, which at first step being of linear order and at next step to be of spatial order.

■

VIBRANT FIELD OF CENTERAL PART OF TRI MONAD

TRI-MONAD FORMAT

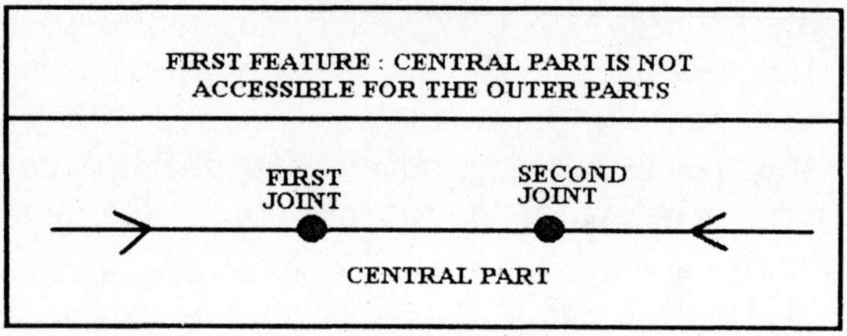

It is this feature of the vibrant field of central part of tri-monad as an Ion (Ion-entity) which is responsible for making pairing of the outer pair of parts as of the order of only the first of pair of pairs of the self-referral field; accepting Ion field pairing as of feature (pairing, pairing).

(From Book : Vedic mathematics Decodes: SPACE BOOK)

■

ASPECT-4.5
THE SPLIT DESERVES TO BE CHASED ALONG BOTH AXES OF SPATIAL ORDER

EXERCISES-1

SPLIT OF DIMENSIONAL FRAME AS FORMAT

1. This and the following two days are to be availed for exercises and expressions for self-evaluation of one's comprehensions about the phenomenon of compactification of origins.
2. One may begin, by first enlisting the technical terms availed for expressions of the concepts.
3. The second exercise which one may undertake would be to draw a flow chart for the split of spirals around the solid quantifiers stem.
4. For this exercise, one may avail the 'split of dimensional frame' as a format.
5. One shall re write upon the first four days topics of the week focusing upon different aspects of fourth feature about compactification of origins.
6. One shall raise as many questions as confront the mind about this phenomenon of compactification of origins and try to reach at their answers.

7. One may, at macro level avail the set up of growth of a plant with focus upon its stem.
8. Further, one may, at micro level, avail the set up around the nucleus of an atom, with nucleus as a stem.
9. One may at insight from the number values generics of words formulations, namely STEM and POLE, being of NVF (STEM) = 57 = NVF (HEIGHT) and NVF (POLE) = 48 = NVF (TREE). Further as that 48=2 × 4 × 6 which is linear equivalence for the dimensional order of 6-space.
10. One shall express one's evaluation of one's comprehension of the phenomenon of compactification of origins as well as also about its different distinct aspects, and the same shall be shared with us.

■

6.2.15 HALF POINT

1. Ultimate is Half Point.
2. Half Point is of the order of INTERVAL.
3. Ultimate makes Half Pairing.

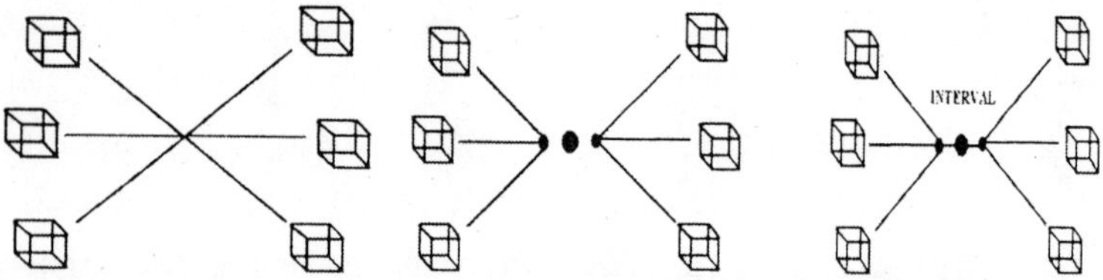

(From Book : Vedic mathematics Decodes: SPACE BOOK)

ASPECT-4.6
THE PAIR OF SPATIAL SPIRALS MAY BE PARALLEL OR CROSS

EXERCISES-2

PARALLEL AND CROSS SPIRALS

FORMAT OF BED OF A FLOWING STREAM

1. One may continue exercises by shifting from the format of '**split of a dimensional frame**' to that of '**bed of a flowing stream**'.
2. Parallel and cross currents formats may be focused for parallel and cross spirals.
3. One shall raise as many questions as confront the mind about this phenomenon of compactification of origins and try to reach at their answers.
4. One shall express one's evaluation of his / her comprehension of the phenomenon of compactification of origins, and their flow formats as spirals around the solid quantifiers stem at its boundary, may those be parallel or cross spiral currents, and the same to be shared with us.

■

ASPECT-4.7
SPIRALS MAY PARALLEL OR CROSS AND THAT TO, TO BE IN IDENTICAL OR OPPOSITE ORIENTATIONS

EXERCISES-3

PARALLEL AND CROSS SPIRALS

FORMAT OF MANIFESTATION LAYERS FOR AXES

1. One may continue exercises by shifting to format of manifestation layers for axes.
2. Parallel and cross currents formats may be focused as parallel and cross manifestations layers.
3. The parallel and cross manifestation layers formats may be availed for chase of split phenomenon of spirals at the boundary of solid quantifiers stem.
4. One shall raise as many questions as confront the mind about this phenomenon of compactification of origins and try to reach at their answers.
5. One shall express one's evaluation of one's comprehension of the phenomenon of compactification of origins, and their flow formats as manifestation layers at boundary of solid quantifiers stem and the same to be shared with us. ■

FEATURE-5

REALITY OF 5-SPACE WITHIN 4-SPACE

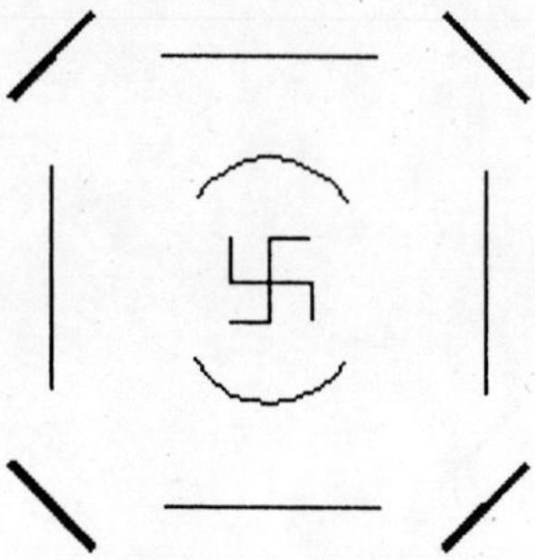

LINKING NOTE

MIDDLE PAIR OF FORTNIGHTS

Features

Third fortnight

1. Feature 05 'About 5-space Reality within 4-space'

2. Feature 06 'About solid quantifiers within 4-space '

Fourth fortnight

3. Feature 07 'About manifestation of solid quantifiers'

4. Feature 08 'About transcendental worlds within Creator's space '

INTRODUCTORY

FIRST PAIR OF FORTNIGHTS

First pair of fortnights features focus upon 'FOUR SPACE REALITY' with concentration upon hyper cube 4, and of it, the feature of utmost interest being its origin fold / transcendental core of origin of 4-space as a seat of compactification of whole range of origins.

This takes us to the Sathapatya foundation. Its symbolic expression may be as under, a set up of Shivligam / transcendental stem / pole installed within Creator's space at its transcendental core.

MIDDLE PAIR OF FORTNIGHTS

This pair of fortnights features focus upon the phenomenon of transcendental worlds within Creator's space. The emergence of field within fields is the distinct features of this set up. It is to be comprehended and chased as the phenomenon of transformation of intelligence field into consciousness field. It may be symbolically depicted as follows as transcendental worlds / hyper cube 5 / 5-space / consciousness field.

5TH FEATURE

ABOUT 5-SPACE REALITY WITHIN 4-SPACE

Fifth feature shall be focusing upon the Reality of real 5-space as Consciousness field emerging within the intelligence field.

This feature may be phased for its chase as follows

1. Creator's space supplies manifestation formats.
2. Each quarter of manifestation format of full manifestation format.
3. Creator's space makes all spaces and their bodies as manifestation layers.
4. Creator's space itself accepts as well accepts manifestation layer feature.
5. Transcendental worlds fulfill Creator's space as solid manifestation layer.
6. The spatial and solid manifestation layers gets superimposed.
7. The phenomenon of superimposed layers lead to simultaneous existence of symmetric and asymmetric set ups.

■

ASPECT-5.1
CREATOR'S SPACE SUPPLIES MANIFESTATION FORMATS

1. This phase and stage of the course takes us to the stage of chase of manifestation formats of Creator's space.
2. 'Idol' of Lord Brahma, Creator the supreme and presiding deity of Creator's space avails the manifestation format of Creator's space for its manifestation.
3. Four Heads of lord are parallel to the four dimensions of 4-space.
4. Each head is equipped with a pair of eyes, parallel to the spatial order of the dimension of 4-space.
5. Lord sits on a lotus seat of eight petals, parallel to eight solid boundary components of hyper cube 4, the representative regular body of 4-space.
6. Lord Brahma meditates in his cavity of heart upon his lord of transcendental worlds, Lord Shiv, five head lord, parallel to the origin of 4-space being the seat of transcendental worlds (5-space).
7. The format of 'Idol' of Lord Brahma, as such emerges to be a set up parallel to that of hyper

cube 4, as of four folds, namely 2-space in the role of dimension, 3-space in the role of boundary, 4-space in the role of domain and 5-space in the role of origin.

8. These features and set up amount to supply of four fold manifestation format by 4-space for manifestation of the whole range of creations / entities.

9. This four-fold set up of manifestation format, as of dimension fold, boundary fold, domain fold, and origin fold, requires availability of four consecutive dimensional spaces contents for manifestation of a dimensional body (domain).

10. In its generality, N as N-dimensional space, shall be manifesting as hyper cube-N as body of N-space (domain) availing four fold manifestation format (N-2, N-1, N, N+1) / (N-2 space as dimension fold, N-1 space as boundary fold, N space as domain fold, N+1 space as origin fold).

11. This four fold features of dimensional body / N-space (domain) may be designated as N-manifestation layer.

12. This designation of N-space (domain) as N-manifestation layer focuses upon the domain fold.

13. Likewise, focus upon dimension fold, boundary fold, and origin fold, as well, when availed, shall be sequencing dimensional bodies. N-space body, with focus upon domain fold, to provide a sequence as of N-domain manifestation layer. However the focus upon boundary fold shall be providing a sequence as of (N-1) boundary manifestation layer.

14. As such we shall be having four distinct sequences

of manifestation layers, viz., (N-2) dimension manifestation layer, (N-1) boundary manifestation layer, N domain manifestation layer and (N+1) origin manifestation layer.

15. The whole range of dimensional spaces in roles of dimensions shall be constituting (N-2) dimension manifestation layers sequence.

16. Likewise, the whole range of dimensional spaces in roles of boundaries shall be constituting (N-1) boundary manifestation layers sequence.

17. The whole range of dimensional spaces in roles of domains shall be constituting (N) domain manifestation layers sequence.

18. And, the whole range of dimensional spaces in roles of origins shall be constituting (N+1) origin manifestation layers sequence.

19. Illustratively, closed interval is a manifestation layer (-1, 0, 1, 2) / (-1-space as dimension, 0-space as boundary, 1-space as domain and 2-space as origin).

20. It would be a blissful exercise to comprehend (-1) space in the role of dimension of 1-space.

21. Further it would also be a blissful exercise to comprehend (0) space in the role of dimension of 2-space.

22. The (N-2) dimension manifestation layers sequence, shall be sequentially taking us to (-1) space as dimension of (+1) space, +1 space as dimension of +3 space, +3 space as dimension of +5 space and so on.

23. Likewise (0) space as dimension of 2-space shall

be at next step would be leading to as 2-space as dimension of 4-space, 4-space as dimension of 6-space and so on.

24. Also (-1) space would be domain with (-3) space as dimension, -3 space shall be a domain (-5) space as dimension and so on.

EXERCISES

1. Chase a sequence of dimensional spaces in the role of dimensions.
2. Chase a sequence of dimensional spaces in the role of boundaries.
3. Chase a sequence of dimensional spaces in the role of domains.
4. Chase a sequence of dimensional spaces in the role of origins.
5. Construct a garland of flowers and see the needle and flowers tagged proceeding on opposite directions parallel to a pair of opposite orientation of a line, as +1 space and –1 space formats, as interconnected (domain, dimension).
6. Observe the play as a setting of points as 0 space in the role of dimension of 2-space.
7. Observe two faces of a plane as settings of (-2) space and (+2) space.
8. Comprehend point as a 0-space domain, as a manifestation layer of four folds with –2 space in the role of dimension, -1 space in the role of boundary, 0-space as domain and +1 space in the role of origin.

■

ASPECT-5.2
EACH QUARTER OF MANIFESTATION FORMAT MAY BE OF EXPRESSION OF FULL MANIFESTATION FORMAT

RECAPITULATION

1. Manifestation format is of four folds; dimension fold, boundary fold, domain fold and origin fold.
2. A set of dimensional contents of four consecutive dimensional spaces simultaneous manifest along the manifestation format, as a manifestation layer of four quarters, as four folds, parallel to four folds of the manifestation format.

Folds of manifestation layer

3. Each fold of the manifestation layer is a set up of a dimensional content of a dimensional space playing the role of such fold of the manifestation layer along the parallel quarter of the manifestation format.
4. The dimensional content of N dimensional space in its dominant role expression as N 'domain' / N body / hyper cube N, in itself, as such, is to play the role of a manifestation layer as of four folds (N-2, N-1, N, N+1).
5. This feature of manifestation makes each quarter

of the manifestation layer being of a manifestation along all the four quarters of a manifestation format, and thereby the quarter of a manifestation layer becoming a four fold manifestation layer.

EXERCISES

1. It would be a blissful exercise to chase the phenomenon of a set of four sequences of manifestation layers constituted by the dimensional spaces four their distinct roles as dimensions, boundaries, domains and origins, on the one hand, and each quarter of manifestation layer emerging to be a full fledged four fold manifestation layer as that each quarter of manifestation format to be of a complete manifestation format.
2. It further would be a blissful exercise to chase this phenomenon of each quarter of manifestation format being a complete manifestation format because of the spatial order of Creator's space / 2-space in the role of dimension of 4-space.
3. Chase domain and boundary of hyper cube 4 as manifestation layers.
4. Chase domain and boundary of cube as manifestation layers.
5. Chase domain and boundary of square as manifestation layers.
6. Chase domain and boundary of interval as manifestation layers.
7. Chase sequence of dimensional spaces in the role of boundaries by jumping from boundary to boundary of hyper cube 4.

8. Chase sequence of dimensional spaces in the role of dimensions by jumping from dimension to dimension of 4-space.
9. Chase sequence of dimensional spaces in the role of origins by jumping from origin to origin of 4-space.
10. Chase sequence of dimensional spaces in the role of N-space domain manifestation layer.
11. Chase sequence of dimensional spaces in the role of (N-1) space boundary manifestation layer.
12. Chase sequence of dimensional spaces in the role of (N-2) space dimension manifestation layer.
13. Chase sequence of dimensional spaces in the role of (N+1) space origin manifestation layer.

■

Glimpse creator-5 as NVF (creation)=85

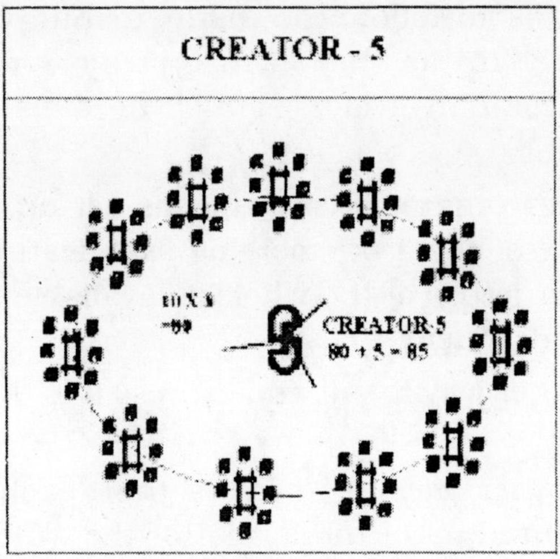

■

ASPECT-5.3
CREATOR'S SPACE MAKES ALL SPACES AND THEIR BODIES JUST BEING THE MANIFESTATION LAYERS

1. Creator's space supplies manifestation format for manifestations as manifestation layers of four folds parallel to four quarters of the manifestation format.
2. As such, all entities / bodies deserve to be chased as manifestation layers.
3. This chase of bodies / entities as manifestation layers is to follow the distinguishing feature of this set up as that each quarter / fold of the manifestation layer, is of full fledged manifestation format.
4. Amongst characteristics of this set up, one most glaring of them is emerging as a feature as that 'the structure of the whole' is available in each of its 'constituent'.
5. The other way to put and express it is that 'part' is 'whole'.
6. Still other way to put it is that 'each point' is structured as of the format of the 'whole body / space'.

7. Illustratively, 'cube' is representative regular body of 3-space as the set up of the cube is of the features of complete 3-space set up.
8. It is because of it that each cube is complete replica of every other cube, big or small.
9. Likewise, is the position of every hyper cube N.
10. This feature of every cube, may be big or small, represents whole of 3-space, deserves to be chased and comprehended well.
11. In this light, the statements like 'N-space' plays the role of boundary of (N+1) space, deserve to be appreciated.
12. 'Solid boundary of 4-space' / '3-space as boundary of 4-space' / '3-space in the role of boundary', and the like expression deserve to be appreciated in the context of the sequence of dimensional spaces in their roles as boundaries of Creator's space.
13. This feature further deserves to be chased by shifting from boundary to boundary but availing the same sequence of dimensional spaces in the roles of boundaries.
14. This sequentially shall be constructing a sequence of boundary of boundary of boundary to any number of steps, and thereby the same sequence of dimensional spaces, in the roles of boundaries, in the process, at its end, shall be reconstructing itself.
15. Likewise, the sequence of dimensional spaces in the roles of dimensions, domains, and origins as well shall be reconstructing themselves.
16. It is this feature of the sequence of dimensional

spaces in its different roles reconstructing itself for manifestations as whole range of sequential steps of any of the quarters of the manifestation layers, which makes the manifestation formats as self referral set ups of spatial order of Creator's space.

EXERCISES

1. Sequentially chase boundary of boundary of boundary order of infinite steps with the help of a sequence of dimensional spaces in the roles of boundaries.
2. Sequentially chase dimension of dimension of dimension order of infinite steps with the help of a sequence of dimensional spaces in the roles of dimensions.
3. Sequentially chase origin of origin of origin order of infinite steps with the help of a sequence of dimensional spaces in the roles of origins.
4. Sadkhas shall permit their mind to transcend to glimpse this phenomenon of sequence of dimensional spaces while in its different roles parallel to the four folds of manifestation layers / manifestation format reconstructing itself parallel to the manifestation of the concerned folds of manifestation layers along concerned quarters of manifestation formats.
5. One shall firstly have an intellectual chase of the boundary of boundary of boundary of hyper cube 6 and then to experientially glimpse the way the sequence of dimensional spaces in the role of boundaries is reconstructed itself by having its

full expression for whole range of set ups of boundaries from –N space to +N space, for all values of N. Here below is reproduced a set up of 960 cubes chasing up till boundary of boundary of boundary of hyper cube 6.

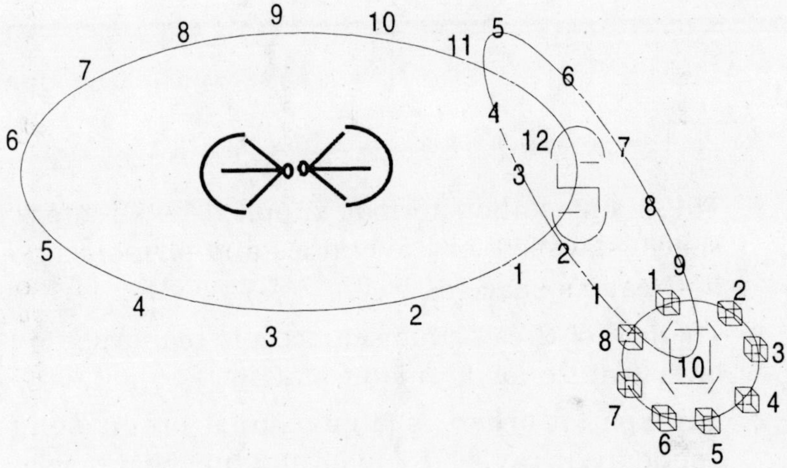

ASPECT - 5.4
CREATOR'S SPACE ITSELF ACCEPTS MANIFESTATION LAYER FEATURE FOR ITS ENTITIES AND PHENOMENON

1. The manifestation format supplied by Creator's space is availed for the entities and phenomenon of Creator's space as well.
2. The Idol of Creator himself accepts manifestation layer feature for its manifestation.
3. The spatial order as dimensional order, solid boundary, hyper solid domain and super hyper solid origin are the specific features of the manifestation layer for the entities and phenomenon of Creator's space.
4. Each of these features has its distinct contribution for the set ups within Creator's space availing manifestation format, but of these the feature of super hyper solid set up for the origin fold has transcendental implication, as that it leads from the manifestation formats to their transcendental bases.
5. The availability of transcendental bases for manifestation formats is the feature, which makes Creator's space as a space of distinction of its

potentialities for creations of whole range of dimensional spaces as manifestation layers.

6. This feature, as of potentialities for creations of dimensional spaces as manifestation layers, is equally extending to Creator's space itself as well, and because of it, it earns the designation of being the Creator's space.
7. The origin fold as transcendental source reservoir makes Creator's space as of transcendental features, as much as that, it has the privilege to get fulfilled with the transcendental features.
8. The scriptures preserve the transcendental phenomenon of the Creator meditating upon His transcendental Lord and getting multiplied ten fold as boundary of transcendental worlds.
9. It is this feature of the domain fold acquiring the role of boundary fold of the origin fold acquiring the role of domain fold, which deserves to be chased and to be comprehended well.
10. Sadkhas shall permit their transcending mind to glimpse this phenomenon of Creator's space acquiring its new role of boundary of the transcendental worlds.
11. In general, this phenomenon deserves to be chased and comprehended as that N-space domain acquiring the role of boundary of (N+1) space.
12. Illustratively, 3-space as domain, acquiring the role of solid boundary of hyper cube 4.
13. Likewise, 2-space as domain, acquires the role of spatial boundary of solids.
14. '3-space' as boundary of hyper cube 4 and further

'3-space' as dimensional order of origin fold of hyper cube 4, are two distinct roles of 3-space, which, as such deserve to be chased along the manifestation format.

15. Further, 3-space bodies having one degree of freedom of motion along fourth dimension, makes the existence phenomenon of manifestation formats as of manifestation layers with its third fold / quarter as of dominant expression of static as well as of dynamic features within Creator's space.

16. As such, the sadkhas shall permit their transcending mind to chase this phenomenon of domain fold of manifestation layer along manifestation format within Creator's space as of static and dynamic features in a domain with solid boundary and also of a solid dimensional order for its origin fold.

To sum up, all the features as of solid boundary, solid dimensional order for the origin fold, third fold of manifestation layer, third quarter of manifestation format, the static and dynamic features of solids within 4-space, are available simultaneously, and as such all these deserve to be chased simultaneously to comprehend the way the manifestation format is availed by the manifestation of entities and phenomenon within Creator's space.

EXERCISES

1. Chase 3-space in the role of boundary of 4-space.
2. Chase 3-space in the role of dimensional order of origin fold of hyper cube 4 / Creator's space.

3. Chase 3-space as format for the dominant expression of body as domain fold.
4. Chase 3-space as format for the domain fold of manifestation format / manifestation layer.
5. Chase 3-space bodies having one degree of freedom of motion within 4-space along its fourth dimension.
6. Chase 4-space entities as manifestation along four fold manifestation format.
7. Chase 4-space phenomenon along manifestation formats as manifestation layers.
8. Chase manifestation of Idol of Creator along manifestation format as manifestation layer of four folds of spatial order, solid boundary, hyper solid domain and super hyper solid origin.

■

Theme Statement

'God Laboratory: Found at Ion'

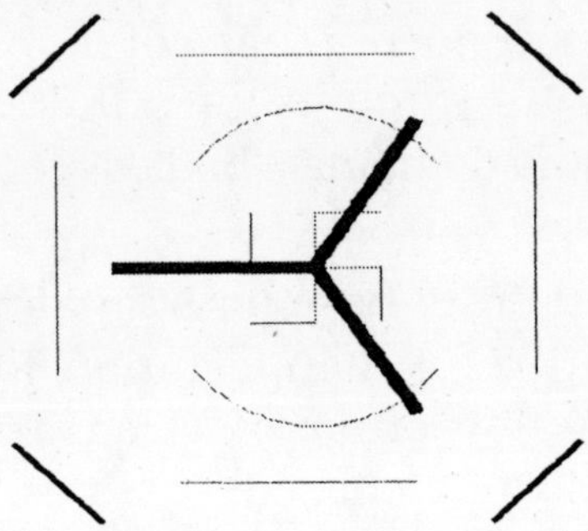

Chase theme statement about **God Laboratory: Found at Ion** geometric base as of **Hyper Cube-4.**

■

ASPECT-5.4
SOLID ORDER OF TRANSCENDENTAL WORLDS FULFILLS CREATOR'S SPACE AS SOLID MANIFESTATION LAYER

1. The fourth fold of manifestation format is the origin fold.
2. The manifestation format for the entities / phenomenon of Creator's space (4-space domain) accepts 5-space in the role of origin fold.
3. '5-space' is a solid order space.
4. '5-space' supplies transcendental base for the manifested domains (4-space domain / bodies).
5. The transcendental base of manifested domains / bodies, as such is of solid order, while the manifested domains / bodies remain to be of spatial order.
6. The solid order base for spatial order base bodies, is a feature of manifestation format for entities and phenomenon of Creator's space, which deserves to be chased.
7. Scripture preserve as that Lord Brahma, Creator the supreme, meditated upon His Lord in cavity of his heart and transcended through the manifested domains and acquired a transcendental role of boundary of transcendental worlds.

8. This phenomenon deserves to be chased and comprehended well by permitting transcending mind to glimpse the feature of transcending to the origin fold and acquiring the features of the transcendental worlds.
9. The transcendence through the manifested domain and attaining of transcendental features, amounts to fulfilling the 4-space domain with solid quantifiers / artifices / features of the solid dimensional order of transcendental worlds at the origin fold.
10. The Creator's space is of spatial order and its supplies manifestation formats for all entities and phenomenon within Creator's space.
11. The emergence of solid quantifiers because of the solid dimensional order released through the origin fold with transcendence from the manifested domain, is the phenomenon as well as the entities (of 3-space feature) which get manifested within Creator's space as of four folds manifestation layers with sold dimensional order, hyper solid 4 boundary, super hyper solid 5 domain and eternal hyper solid 6 origin.
12. This amounts to super imposition of solid quantifiers set ups upon spatial quantifiers set up of Creator's space.
13. This phenomenon of emergence of solid quantifiers within Creator's space deserves to be chased fully.
14. This happening of emergence of solid quantifiers within Creator's space amounts to the simultaneous availability of spatial and solid orders.

15. This happening of emergence and manifestation of solid quantifiers within Creator's space amounts to availability of manifestation layers (3, 4, 5, 6) of features of transcendental worlds within Creator's space.
16. Therefore the sadkhas shall permit the transcending mind to chase this phenomenon as emergence of transcendental worlds within Creator's space.

EXERCISES

1. Chase origin of hyper cube 4 as a seat of transcendental worlds (5-space).
2. Chase origin fold of hyper cube 4 as of solid dimensional order.
3. Chase hyper cube 5 as a manifestation layer (3, 4, 5, 6).
4. Chase the phenomenon of emergence of solid quantifiers within Creator's space.
5. Chase the phenomenon of manifestation of solid quantifiers within Creator's space.
6. Chase the phenomenon of super imposition of solid quantifiers upon spatial quantifiers.
7. Chase the phenomenon of simultaneous availability of spatial order and solid order within Creator's space.
8. Chase the phenomenon of emergence of transcendental worlds within Creator's space.

■

ASPECT-5.5
THE SPATIAL AND SOLID MANIFESTATION LAYERS GETS SUPERIMPOSED UPON EACH OTHER

1. The phenomenon of spatial and solid orders being simultaneous available within Creator's space deserve to be chased.
2. The spatial order is of Creator's space.
3. The spatial order may be chased as a spatial manifestation layer (2, 3, 4, 5) with focus upon the role of 2-space as dimension.
4. The solid order is of transcendental worlds.
5. The solid order may be chased as a solid manifestation layer (3, 4, 5, 6) with focus upon the role of 3-space as dimension.
6. The emergence of solid quantifiers within hyper cube 4 because of the solid dimensional order of transcendental worlds as origin fold of hyper cube 4 amounts to the emergence of solid manifestation layers within hyper cube 4.
7. Hyper cube 4 itself being of spatial order, as such, it inherently has the supply of spatial manifestation layers within its domain.
8. The emergence of solid manifestation layers within

domain of hyper cube 4 amounts to super imposition their of upon the spatial manifestation layers.

9. This phenomenon of superimposition of solid manifestation layer upon spatial manifestation layers deserves to be chased as the phenomenon of transcendental carriers.
10. The solid quantifiers supplied by the solid dimensional order of transcendental worlds fulfills Creator's space with transcendental features and the same when super imposed upon the spatial manifestation layers, the same transforms them into transcendental carriers.
11. The phenomenon of transition and transformation from spatial manifestation layers into transcendental carriers is the phenomenon of acquisition of transcendental features by the spatial order of the Creator's space.
12. Scriptures well preserve the chase of this phenomenon from the phase and stage of transformation of Shad Chakras format (6 eternal circuits) format of human body to Dash Chakras (10 eternal circuits) format of human body, which along artifices 6 and 10 and parallel to it 6 boundary component of 3-space and 10 boundary component of 5-space is a transition and transformation from solid order to transcendental worlds.
13. Scriptures further well preserve the chase of riding of impulses of consciousness upon the bits of intelligence / being riding the transcendental carriers within sunlight, and this phenomenon of transcendental carriers deserves to be chased as

superimposition of solid quantifiers upon spatial quantifiers within Creator's space.

14. Sadkhas shall permit their transcending mind to glimpse this phenomenon of transcendental carriers within Creator's space as phenomenon of solid quantifiers manifesting as solid manifestation layers and getting superimposed upon spatial quantifiers as spatial manifestation layers.
15. Sadkhas shall further permit their transcending mind to glimpse this phenomenon of transcendental carriers of Being taking off from the human body and riding the transcendental carriers within sun rays parallel to the phenomenon of super imposition of solid manifestation layers upon spatial manifestation layers within Creator's space.

EXERCISES

1. Chase the phenomenon of emergence of solid quantifiers and manifesting as solid manifestation layer within Creator's space.
2. Chase the phenomenon of solid manifestation layers super imposing upon spatial manifestation layers.
3. Chase the phenomenon of transcendental carriers within Creator's space as the phenomenon of transcendental features acquired by the spatial manifestation layers of Creator's space.
4. Chase the phenomenon of the transcendental carriers within rays of the sun.

■

ASPECT-5.7

THE PHENOMENON OF EMERGENCE OF TRANSCENDENTAL WORLDS WITHIN CREATOR'S SPACE DESERVES TO BE CHASED AS A PHENOMENON OF SUPERIMPOSED LAYERS, AND AS A CONSEQUENCE THERE OF, BOTH SYMMETRIC AND ASYMMETRIC SET UPS CO-EXIST.

1. The spatial order is of symmetric features as 2=1+1. The solid order is of symmetric features as well as of asymmetric features 3=1+1+1=2+1.
3. The phenomenon of super imposition of solid order upon spatial order is the phenomenon of both symmetric and asymmetric set ups simultaneously available.
4. The scripture preserve descriptions of symmetric and asymmetric existence phenomenon of transcendental worlds, as much as that, the beings there may be of 'trunk' only, eyes in the belly, no arms, many arms, asymmetric arms, one leg, many legs, flat face, mouth on the top of the head, in belly, many heads, and so on.
5. This chase of sold order super imposed upon spatial order may be had by simultaneously availing 'square' and 'cube'. It may be cube 'put' on a square plate. It may be a square plate put on a cube.
8. It may be a synthetic set up with one part as a square and other part as a solid.

9. It may be a square and cube within a cube or cube and a square within hyper cube 4 or within hyper cube 5.
10. It may be 2-space as dimension and 3-space as origin; 2-space as dimension and 3-space as domain; 2-space and 3-space both being dimensions or boundaries or domains or origins.
11. It may be 2-space and 3-space as paired dimensional frames of zero order and linear order.
12. It may be 2-space and 3-space as paired artifices (2 and 3).
13. Like that, it would be a blissful exercise for the sadkhas to chase this phenomenon of simultaneous availability of spatial and solid orders within Creator's space with transcendental base.

CHASE EXERCISE

8. Creator's space supplies manifestation formats.
9. Each quarter of manifestation format of full manifestation format.
10. Creator's space makes all spaces and their bodies as manifestation layers.
11. Creator's space itself accepts as well accepts manifestation layer feature.
12. Transcendental worlds fulfill Creator's space as solid manifestation layer.
13. The spatial and solid manifestation layers gets superimposed.
14. The phenomenon of superimposed layers lead to simultaneous existence of symmetric and asymmetric set ups. ■

FEATURE-6

5-SPACE DESERVES TO BE CHASED AS SPACE SUPPLYING FIVE STEP LONG MEASURING ROD

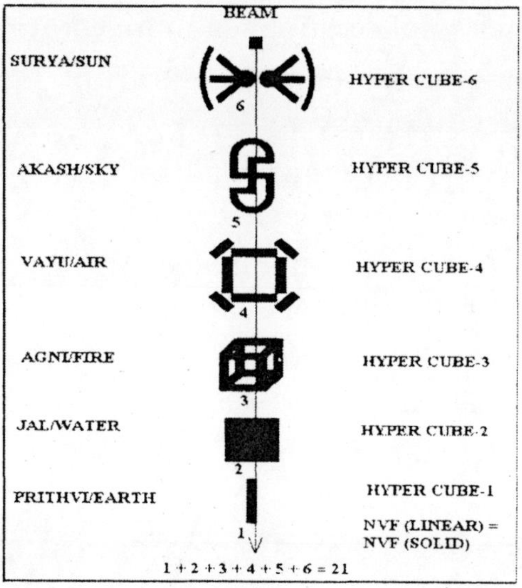

ASPECT-6.1

5-SPACE DESERVES TO BE CHASED AS SPACE SUPPLYING FIVE STEP LONG MEASURING ROD

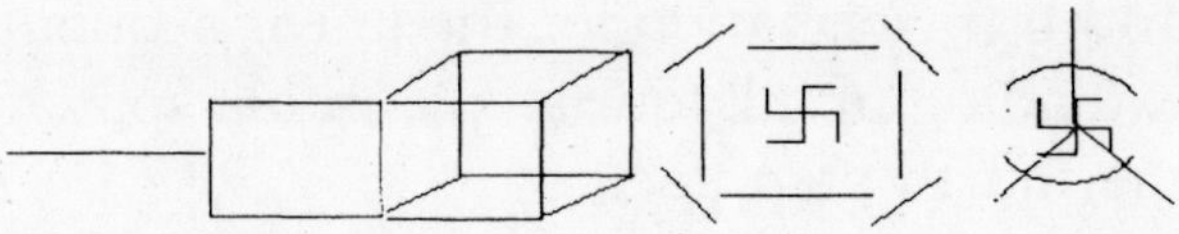

1. Hyper cube 4 (□) is a manifestation layer (2, 3, 4, 5).
2. Fourth fold is the origin fold; it is origin source reservoir of transcendental features, a transcendental base (卐) of manifested domain (□).
3. The transcendental base (卐), as origin source reservoir of transcendental features is the transcendental world of solid dimensional order.
4. The transcendental base (卐), being the transcendental world / 5-space, it accepts a measuring rod constituted by representative regular bodies of 1 to 5 space, that is, interval, square, cube, hyper cube 4 and hyper cube 5.
5. This synthetic set up avails sequential range of five steps of the sequence of dimensional spaces as dimensions, boundaries, domains, origins and transcendence ranges.

6. Of these, the first four, namely the sequential ranges of sequence of dimensional spaces in the roles of dimensions, boundaries, domains and origins are availing four fold manifestation formats manifesting manifestation layers of four quarters.
7. The fifth sequential step is of transcendence range, transiting and transforming the manifestation format into a transcendence range at the origin fold because of the availability of spatial order within Creator's space.
8. The fifth sequential step, as such attains transition and transformation for the linear sequential order extending up till four steps, into a spatial order at the fourth step itself.
9. The spatial order of the fourth step, in a sequence leads ahead to the next fifth step as of a solid order.
10. This range, as of five steps, attaining solid order, is a transcendence range of five folds, leading to the transcendental worlds parallel to five steps long measuring rod as a synthetic set up of representative regular bodies of 1 to 5 space.
11. This way, the origin fold as seat of transcendental worlds accepts a measuring rod of five steps long range attaining transcendence to the transcendental base of the manifested domains within Creator's space.
12. The transcendental base, being origin source reservoir of transcendental features of transcendental worlds / 5-space, fulfills the Creator's space / 4-space domain with transcendental features of solid quantifiers which

manifest as solid manifestation layers and get superimposed upon the spatial manifestation layers of dimensional order of Creator's space.

13. Sadkhas shall permit their mind to transcend availing five steps long transcendence range and glimpse the phenomenon of transcendental base of manifested domains within Creator's space.

14. Sadkhas shall permit the transcending mind to glimpse its transcendence paths of five folds of five sequential steps of five steps long each of the sequence of dimensional spaces in distinct five roles of dimensions, boundaries, domains, origins and as transcendence ranges.

15. Sadkhas shall permit the transcending mind to glimpse the sequential progression of transcendence along the transcendence range up till first four steps parallel to four quarters of manifestation layers along four fold manifestation format and then to attain transition and transformation from spatial order of Creator's space to solid order of transcendental worlds.

16. Sadkhas shall permit the transcending mind to glimpse the sequential progression of first four step of the transcendence range parallel to four fold manifestation layers of manifestation format with each quarter of the format being a full fledged format manifesting complete manifestation layers and their by structuring a spatial order, and then at the fifth step to attain transition and transformation of solid order, not only for the fifth fold but also for all the five folds as 1 to 5 space / hyper cubes 1 to 5 within 5-space.

17. Sadkhas shall experience the transcending mind creating intelligence field and also at the fifth step attaining transition and transformation for the intelligence field into consciousness field.
18. Sadkhas shall experience the emergence of five step long transcendence range for four fold manifestation layers.

EXERCISES

Write short notes on the following conceptual terms.

a. Manifestation format
b. Four fold manifestation format
c. Four quarters of manifestation format
d. Manifestation layer
e. Four folds of manifestation layers
f. Four quarters of manifestation layers
g. Measuring rod
h. Measuring rod of 4-space
i. Measuring rod of 5-space
j. Transcendence range
k. Five fold transcendence range.
l. Transcendental worlds
m. Transcendental features
n. Solid order
o. Solid order transcendental worlds
p. Five space
q. Fifth step of transcendence range.
r. Origin source reservoir of transcendental features
s. Transcendental base of manifested domain

And all other technical / conceptual terms ■

ASPECT-6.2
5-SPACE DESERVES TO BE CHASED AS SPACE SUPPLYING FIVE STEP LONG TRANSCENDENCE RANGE FOR FOUR FOLD MANIFESTATION LAYERS

RECAPITULATION

It may be recapitulated that 4-space accepts chase as four steps long manifestation format.

5-SPACE

1. The 5-space is a transcendental world.
2. It accepts a five steps long measuring rod.
3. It is a solid order world.
4. It provides transcendental base to manifestation bodies of Creator's space.
5. It accepts transcendence to it from the manifested domains of Creator's space.
6. This transcendence step is in continuity of the four steps long manifestation format supplied by the Creator's space.
7. This, this way works out a five steps long transcendental range as a transcendence path leading from Creator's space to transcendental worlds.
8. As such, sadkhas may accept 5-space as the space

supplying five step long transcendence range for four fold manifestation layers.

9. The transcendental world being a solid order world, it with its transcendence process transcends through the manifested domain (3-space) of its dimensions (solid order) and attains Creator's space as base for its dimensional order.

10. The sadkhas fulfilled with an intensity of urge to fully know and to completely chase the phenomenon of emergence of transcendental features within Creator's space and the phenomenon of emergence of Creator's space as base for the dimensional order of the transcendental worlds shall permit their transcending mind to go in trans time and again and firstly to perfect their intelligence in respect of the emergence of intelligence field itself during transcendence of the transcending mind, and secondly in respect of the emergence of the consciousness field within the intelligence field with the transcending mind glimpsing the inner folds and inner most fold of the transcendental world as self referral domain being 6-space accepting 4-space as its dimensional order.

11. Sadkhas shall also intellectually chase the different roles of 1 to 5 space of the transcendence range, which shall be of the following formats:

    Formats of transcendence range for different roles of 1-space

    -3, -2, -1, 0, 1

    -2, -1, 0, 1, 2

    -1, 0, 1, 2, 3

0, 1, 2, 3, 4

1, 2, 3, 4, 5

Formats of transcendence range for different roles of 2-space

-2, -1, 0, 1, 2

-1, 0, 1, 2, 3

0, 1, 2, 3, 4

1, 2, 3, 4, 5

2, 3, 4, 5, 6

Formats of transcendence range for different roles of 3-space

-1, 0, 1, 2, 3

0, 1, 2, 3, 4

1, 2, 3, 4, 5

2, 3, 4, 5, 6

3, 4, 5, 6, 7

Formats of transcendence range for different roles of 4-space

0, 1, 2, 3, 4

1, 2, 3, 4, 5

2, 3, 4, 5, 6

3, 4, 5, 6, 7

4, 5, 6, 7, 8

Formats of transcendence range for different roles of 5-space

1, 2, 3, 4, 5

2, 3, 4, 5, 6

3, 4, 5, 6, 7

4, 5, 6, 7, 8

5, 6, 7, 8, 9

12. Sadkhas shall experience the different roles of Space 1 to 5 space as formats of different folds of the transcendence range.

EXERCISES

1. Express about the role of 1-space as first fold of the transcendence range.
2. Express about the role of 1-space as second fold of the transcendence range.
3. Express about the role of 1-space as third fold of the transcendence range.
4. Express about the role of 1-space as fourth fold of the transcendence range.
5. Express about the role of 1-space as fifth fold of the transcendence range.
6. Express about the role of 2-space as first fold of the transcendence range.
7. Express about the role of 2-space as second fold of the transcendence range.
8. Express about the role of 2-space as third fold of the transcendence range.
9. Express about the role of 2-space as fourth fold of the transcendence range.
10. Express about the role of 2-space as fifth fold of the transcendence range.
11. Express about the role of 3-space as first fold of the transcendence range.
12. Express about the role of 3-space as second fold of the transcendence range.

13. Express about the role of 3-space as third fold of the transcendence range.
14. Express about the role of 3-space as fourth fold of the transcendence range.
15. Express about the role of 3-space as fifth fold of the transcendence range.
16. Express about the role of 4-space as first fold of the transcendence range.
17. Express about the role of 4-space as second fold of the transcendence range.
18. Express about the role of 4-space as third fold of the transcendence range.
19. Express about the role of 4-space as fourth fold of the transcendence range.
20. Express about the role of 4-space as fifth fold of the transcendence range.
21. Express about the role of 5-space as first fold of the transcendence range.
22. Express about the role of 5-space as second fold of the transcendence range.
23. Express about the role of 5-space as third fold of the transcendence range.
24. Express about the role of 5-space as fourth fold of the transcendence range.
25. Express about the role of 5-space as fifth fold of the transcendence range.

■

ASPECT - 6.3
EACH OF FIVE STEPS OF TRANSCENDENCE RANGE TO ACCEPT FULL TRANSCENDENCE RANGE FOR TRANSCENDENCE THERE FORM

1. '1' as artifice 1 is parallel to '1-space', '1-manifestation layer' and is also parallel to '1 transcendence range'.
2. The 5 x 5 grid, as a set up, is of value 5 x 5 =25.

   1, 1, 1, 1, 1

   1, 1, 1, 1, 1

   1, 1, 1, 1, 1

   1, 1, 1, 1, 1

   1, 1, 1, 1, 1

   5 × 5=25
3. The sequential chase of this 5 × 5 grid leads to a cubic value 5 × 5 × 5

   1, 2, 3, 4, 5

   2, 3, 4, 5, 6

   3, 4, 5, 6, 7

   4, 5, 6, 7, 8

   5, 6, 7, 8, 9

   5 × 5 × 5 =125

4. This sequential chase along a spatial format resulting into a solid / cubic value has geometric message as that, the transcendental worlds / 5-space / artifice 5 / transcendence range, as of sequential progression, is parallel to the measuring rod of 5 steps long synthetic set ups of sequence of dimensional spaces of five steps long ranges have the characteristics features of five distinct roles for each dimensional space, which in other words means that each of the five folds of the transcendence range, in fact is a full fledged transcendence range.
5. The sadkhas shall chase this characteristics features of this set ups of dimensional spaces manifesting as domains with transcendental base and their by making available a transcendence from manifested domain to transcendental base with each fold of transcendence range being a full fledged transcendence range.
6. It is here where the intellectual chase which is intelligence field based processing system shall be needing argumentations from the experiential chase which is consciousness field based processing system.
7. It is here where the geometric formats give place to artifices of numbers for processing beyond the transcendental worlds as self referral domains of hyper solid dimensional orders.
8. The sadkhas shall sequentially attain perfection for handling linear order, spatial order, solid order and then hyper solid orders.
9. The solid order makes lively the transcendental

worlds.

10. The hyper solid 4 order makes lively the self referral domain of 6-space.
11. The hyper solid 5 order makes lively the unity state domains of 7-space.
12. The hyper solid 6 order makes lively the automation state domains of 8-space.
13. The hyper solid 7 order makes lively the ultimate affine Brahman domains as 9-space.

EXERCISES

1. Chase within Creator's space the hyper cubes 1 to 9 as manifestation layers 1 to 9, viz., closed interval, square, cube, hyper cubes 4 to 9.
2. Chase hyper cubes 1 to 9 as four fold manifestation layers along manifestation format.
3. Chase hyper cubes 1 to 9 in terms of their measuring rods.
4. Chase hyper cubes 1 to 9 in terms of transcendence ranges of 5 steps long set ups of formats of dimension fold, boundary fold, domain fold, origin fold, and transcendence range.
5. Chase each of the four steps of the manifestation format as full fledged formats.
6. Chase each of the five steps of the transcendence range as full fledged transcendence ranges.

■

ASPECT-6.4
THE DIMENSIONAL ORDER OF EACH STEP OF TRANSCENDENTAL RANGE AS WELL BEING OF FULL TRANSCENDENCE RANGE

1. Construct a 6 x 6 grid.
2. Construct another 5 x 5 grid with centers of grid zones of 6 x 6 grid as grid joints for 5 x 5 grid.
3. Superimpose 5 × 5 grid upon 6 × 6 grid.
4. The grid joints of 6 x 6 grid would get super imposed with centers of the grid zones of 5 x 5 grid.
5. This way the centers of grid zones of 6 x 6 grid would get super imposed by the grid joints of 5 x 5 grid, and on the other hand the centers of grid zoom of 5 x 5 grid would get super imposed by the grid joints of 6 x 6 grid.
6. Accepting centers of grid zones as origin source reservoir seats, there would flow into grid zones the quantifiers of the dimensional order of the origin fold / base of grid zones.
7. The grid zones being 'spatial' / 2-space, with 0 space in the role of dimension, its origin fold seat would be of 3-space with 1-space in the role of dimension thereof.
8. As such the linear order of 3-space / linear

quantifiers would get fulfilled within the grid zones through their centers.

9. The linear quantifiers flowing into grid zones at their centers, because of the spatial set up of the grid zones, shall be manifesting as circumferences of circles around the center of the grid zones.
10. The linear order / 1-space / 1-manifestation layer (-1, 0, 1, 2) shall be accepting reversal of orientations for its ascendance to be transcendence, and thereby the transcendence range (-1, 0, 1, 2, 3) shall be taking to dimension of dimension level, with transcendence from the middle to the end below and simultaneously there being ascendance from the middle to the end upward.
11. This, this way emerges to be a full transcendence range being available for its every step of the range of transcendence, parallel to the above two fold transcendence-ascendance flow from the middle, and this feature being not only restricted for any particular fold of manifestation format, and in particular, it shall also be available for the dimension fold as well.
12. It would be a blissful exercise to chase this phenomenon of full transcendence range being available for each of the five steps of the transcendence range, with specific focus at this stage, as to be upon the dimension fold.

■

ASPECT - 6.5
THE BOUNDARY FOLD OF EACH STEP OF TRANSCENDENTAL RANGE AS WELL BEING OF FULL TRANSCENDENCE RANGE

1. One shall be through the chase steps of day 39 dated 23-10-2005 by taking as that now instead of dimension fold, focus is upon the boundary fold.
2. It would be a blissful exercise to chase this phenomenon of full transcendence range being available for each of the five steps of the transcendence range, with specific focus at this stage, as to be upon the boundary fold.

■

Chase transcendence along the format of a measuring rod

MEASURING ROD

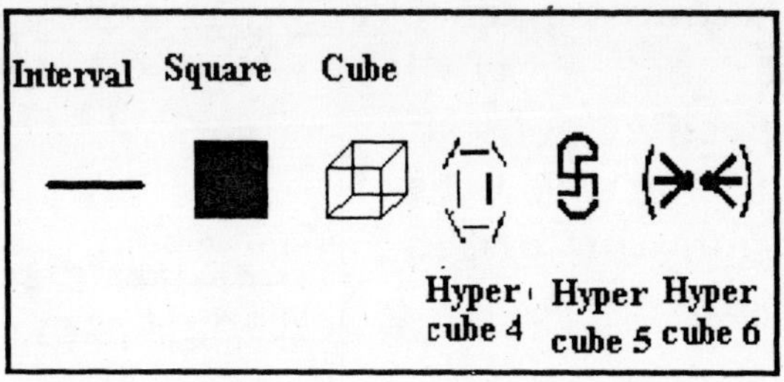

■

ASPECT-6.6
THE ORIGIN FOLD OF EACH STEP OF TRANSCENDENTAL RANGE AS WELL BEING OF FULL TRANSCENDENCE RANGE

1. One shall be through the chase steps of day 39 dated 23-10-2005 by taking as that now instead of dimension fold, focus is upon the origin fold.
2. It would be a blissful exercise to chase this phenomenon of full transcendence range of a beam within space, being available for each of the five steps of the transcendence range, with specific focus at this stage, as to be upon the origin fold.

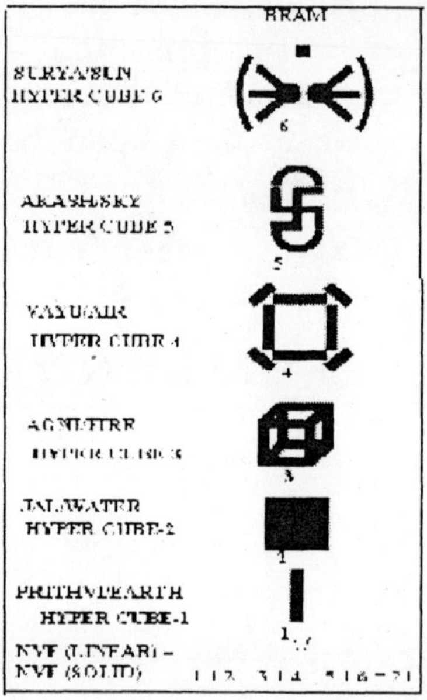

■

ASPECT-6.7
THE PHENOMENON OF FIRM ESTABLISHMENT OF SOLID QUANTIFIERS WITHIN CREATOR'S SPACE DESERVES TO BE CHASED FULLY

1. Creator's space is of spatial order.
2. Transcendental worlds are of solid order.
3. Solid order accepts solid quantifiers.
4. The origin source reservoir of Creator's space is the solid order transcendental worlds.
5. The origin source reservoir of the Creator's space being of solid order, so the 4-space domain gets fulfilled with solid quantifiers.
6. The solid quantifiers get four fold spatial order format for its manifestation.
7. This four fold manifestation of solid quantifiers is of the set up of solids / cubes / spheres as bodies of 3-space with 4-space as origin fold.
8. The solid quantifiers manifested along the manifestation format as cubes solids acquire one degree freedom of motion within Creator's space because of the fourth dimension available for motion.
9. The Creator's space being of spatial order, as such

it fulfills solid domains / cubes with spatial quantifiers.

10. The spatial quantifiers provide spatial format for linear dimensional orders of solids-domains, and as such the linear dimensions accept formations as a pair of half dimensions.
11. The pair of half dimensions within spatial set up emerges to be of 'artifice-1' as half of 'artifice-2', and this way the half dimensions in fact emerge to be of 'full linear order' and thereby the three dimensional frame multiplies as a pair of three dimensional frames.
12. It is this phenomenon of a three dimensional frame multiplying as a pair do three dimensional frame which deserve to be chased as firm establishment of solid quantifiers within Creator's space.

FINAL EXERCISE

As such, in the light of above features, it may be taken as a final exercise to chase phenomenon of firm establishment of solid quantifiers within Creator's space.

Sathapatya foundation ■

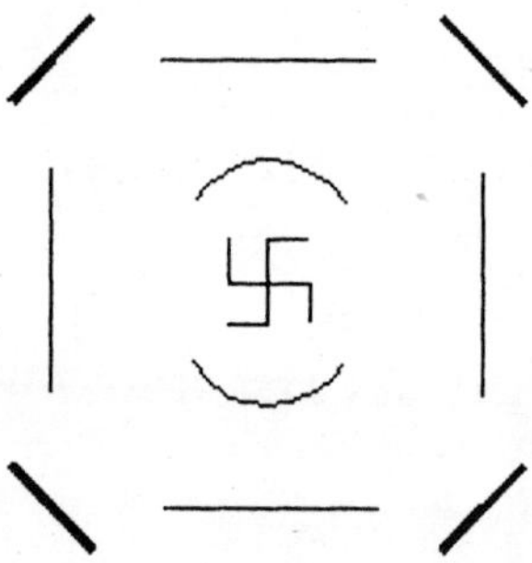

■

FEATURE-7
MANIFESTATIONS OF SOLID QUANTIFIERS

FEATURE-7

Seventh feature of this chase is going to be about the manifestations of solid quantifiers and the transcendental range sealed at the origin fold.

ASPECT-7.1
CHASE REAL 5-SPACE WITHIN CREATOR'S SPACE AS CONSCIOUSNESS FIELD WITHIN INTELLIGENCE FIELD

INTRODUCTORY

1. Here we are reaching a 'phase and stage' of the course, where we are concentrating upon the 'reality of phenomenon happenings availing geometric formats' for whose comprehension and understanding this initial stage course is attempting exposure for the mind for its own chase of creation of intelligence and consciousness field while transcending through mundane / 3-space and manifested / 4-space in the roles of boundaries.
2. One such phenomenon happening is of the intelligence waves carrying the manifested values.

3. This phenomenon happening is essentially of the

Creator's space, which is enveloped within solid / mundane / manifested bodies domains.

RECAPITULATION

4. Solid boundary of Creator's space emerges as of a manifestation process, which accepts solid quantifiers from its origin source fold of transcendence values of transcendental worlds (5-space, which is of solid order).
5. 'Mind', while transcends through this solid boundary, releases the manifested values, which have been manifesting there in the role of the boundary.

PROJECTIONS

6. One of the features of these values during their manifestation and acquiring of the role of is the 'boundary placement'.
7. This feature is of characteristics 'of riding a domain'.
8. It is this characteristics which equips the 'manifested values' to ride the intelligence waves which are of spatial order while the manifested values are simply of a linear order of the solid boundary.

PART-A

MANIFESTED VALUES

9. These values (manifested values) are supplied by the solid boundary of Creator's space.
10. Sadkhas shall first chase, as to how, the solid quantifiers, from there initial stage of emergence

from origin source fold of Creator's space, fulfill the Creator's space, manifest a solid manifestation layer, vest solids within Creator's space domain the potentialities of a dynamic solids within the domain and as of static solids at the boundary of the Creator's space.

11. Sadkhas shall permit their transcending mind to glimpse the way it transcends through the manifested boundaries of the Creator's space.
12. Sadkhas shall also glimpse the happening of the transcendence through the manifested boundary being possible because of the manifestation format and set up accepting a transcendental base originally supplied by the 'origin source fold of the transcendental values' which were supplied as solid quantifiers within the Creator's space.
13. Sadkhas shall also glimpse the release of the manifested values while the mind transcends through the manifested boundary of Creator's space.
14. These values run parallel to the order / dimensional order of the solid boundary.
15. These values of dimensional order are the measures of the measuring rods, and as such '1-space being the dimensional order of solid boundary' is the measure of the measuring rod of 3-space set up of solids as boundary of Creator's space.
16. Sathapatya Ved enlightens as that Lord Brahma is the presiding deity of the measure while Lord Vishnu is the presiding deity of the measuring rods.

17. Lord Brahms, creator the supreme, is a four head lord, and avails manifestation format for manifestation for its Idol and also for entities as well as for phenomenon happenings within Creator's space.
18. With this the '1-space' as dimensional order would be a manifestation format, which may be manifesting as 1-manifestation layer (-1, 0, 1, 2).
19. As such, the manifested values shall be flowing out of solid boundary as 1-manifestation layer.
20. This flow of 1-manifestation layer while mind transcends through it, shall be entering Creator's space-domain, which is of a spatial order manifesting as 2-manifestation layer (0, 1, 2, 3).
21. The 'riding characteristics' of dynamic solids, previously resting at boundary, shall be inherently equipping the released manifested values with the potentialities (to ride the spatial order of 2-manifestation layer of intelligence field of Creator's space.

PART-B

INTELLIGENCE FIELD

22. It would be a blissful exercise for the sadkhas to chase their transcending mind creating intelligence field while transcending through the manifested boundary of the Creator's space.
23. It further would be a blissful exercise for the sadkhas to chase their transcending mind creating consciousness field within intelligence field transcending through the manifested hyper solid boundary of transcendental worlds.

24. Still further, it would be a blissful exercise for the sadkhas to chase their transcending mind even transcending through the transcendental boundary of self referral domains and in the process transforming itself as 'Self Referral Being' as life virtues emerging within the consciousness field and taking of for riding the transcendental carriers within rays of the sun / 6-space domain.

■

Experiential Bliss

1. Vedic comprehension is that there is absolute domain/format being designated as *Brahman* domain/*Nav Braham*/9-Space.
2. Within transcendental world it is reached on transcendental format of five folds as (5, 6,7,8,9).
3. Within creator's Space it is reached as a manifestation layer of four folds (6,7,8,9) and 6-Space/*Vishnu Lok*/*Atman*/Sun is reached as a manifestation layer (3,4,5,6).
4. The transcendental world with transcendental format of five folds permits transition from one manifestation layer to next manifestation layer. Illustratively from manifestation layer (1,2,3,4) to (2,3,4,5).
5. This transition through transcendence from creator's Space to transcendental world is availed by the Vedic systems for transition from old formats to new formats.

■

ASPECT-7.2
CHASE CONSCIOUSNESS FIELD WITHIN INTELLIGENCE FIELD

RECAPITULATION

1. It may be recapitulated as that the transcending mind creates intelligence field with its transcendence through the solid boundary and entering the Creator's space domain.
2. It may further be recapitulated as with further transcendence of the transcending mind through hyper solid boundary of the transcendental worlds and entering the transcendental worlds, a consciousness field stands creating within the intelligence field.

Consciousness Field Within Intelligence Field

3. The phenomenon of emergence of consciousness field within intelligence field is the transcendence phenomenon happening of emergence of transcendental features within Creator's space.
4. The geometric format of this transcendence phenomenon happening is of the solid quantifiers fulfilling the Creator's space and manifesting therein as solid manifestation layer superimposed

upon the spatial manifestation layer of the spatial order of the Creator's space.

5. As such the transcendence phenomenon happening of emergence of consciousness field within Creator's space deserves to be chased as solid dimensional order superimposed upon the spatial dimensional order.

EXERCISES

1. Chase the order as manifestation layer

   (a) of Creator's space .

   (b) of transcendental worlds.

2. Chase the phenomenon of solid quantifiers

   (a) fulfilling the Creator's space-domain.

   (b) manifesting wihtin creator's space.

   (c) solid order superimposing spatial order

6. Chase the emergence of intelligence field within Creator's space.

7. Chase the emergence of consciousness field within transcendental worlds.

8. Chase the emergence of consciousness field within intelligence field.

9. Chase the consciousness field as of solid order and as solid manifestation layer.

10. Chase the consciousness field as of solid order and as solid manifestation layer superimposed upon the spatial manifestation layer of spatial order of intelligence field within Creator's space.

■

ASPECT-7.3
CHASE THIS PHENOMENON AS CONSCIOUSNESS IMPULSES SUPERIMPOSED UPON INTELLECTUAL WAVES

RECAPITULATION

1. Recapitulate the consciousness field as of solid order and as solid manifestation layer superimposed upon the spatial manifestation layer of spatial order of intelligence field within Creator's space.

CONSCIOUSNESS FIELD

2. The consciousness field is of solid dimensional order which manifests within the Creator's space as a solid manifestation layer.
3. The Creator's space being of spatial order manifesting as a spatial manifestation layer, it permits superimposition of a solid manifestation layer for carrying the linear dimensional order values of the solid order.
4. The parallel geometric format is available to the impulses of consciousness to be carried by the intelligence waves.
5. It would be a blissful exercise for the sadkhas to

chase the phenomenon of consciousness impulses superimposing upon intelligence waves.

6. It further would be a blissful exercise for the sadkhas to intellectually validate their experiences of the transcending mind glimpsing this happening while transcending through the inner layers of the transcendental worlds emerging within Creator's space.

EXERCISES

1. Chase the transcendence phenomenon
   - (a) of consciousness impulses of waking state getting superimposed upon the intelligence waves within Creator's space.
   - (b) consciousness impulses of dream state getting superimposed upon the intelligence waves within Creator's space.
3. Chase the transcendence phenomenon of consciousness impulses of deep sleep state getting superimposed upon the intelligence waves within Creator's space as a linear manifestation layer of solid format superimposing upon the spatial manifestation layer of hyper solid format.
4. Chase the transcendence phenomenon of consciousness impulses of transcendental state getting superimposed upon the intelligence waves within Creator's space as a linear manifestation layer of hyper solid format superimposing upon the spatial manifestation layer of super hyper solid format.

■

ASPECT-7.4
CHASE INTELLECTUAL WAVES AS CARRIERS OF MANIFESTED VALES RECAPITULATION

1. It may be recapitulated that the intelligence field is created within a Creator's space domain by the transcending mind by transcending through the manifested boundary.
2. It may be further recapitulated that the manifested boundary is linear order solid boundary while the intelligence field is availing the spatial order of Creator's space domain.

MANIFESTED VALUES AND INTELLIGENCE WAVES

3. The dimensional orders of the boundary and domain of Creator's space, being of linear and spatial features, as such the manifested values and intelligence waves as well are to be of linear and spatial formats.
4. As such the manifested values are carried by the intelligence waves as linear order is carried by the spatial order.

EXERCISES

1. It shall be a blissful exercise for the sadkhas to chase their transcending minds transcending

through manifested boundaries and creating intelligence field.

2. It shall further be a blissful exercise for the sadkhas to chase the way the manifested values are carried by the intelligence waves.
3. One shall express fully and to completely evaluate one's comprehension and understanding of the transcendence phenomenon happening of creation of intelligence field by the transcending mind.
4. One shall express fully and to completely evaluate one's comprehension and understanding of the transcendence phenomenon happening of the intelligence manifesting as intelligence waves within Creator's space.
5. One shall express fully and to completely evaluate one's comprehension and understanding of the transcendence phenomenon happening of the manifested values being released by the manifested boundary with transcendence of the mind through it.
6. One shall express fully and to completely evaluate one's comprehension and understanding of the transcendence phenomenon happening of the transcendental world manifested values being carried by the intelligence waves.

■

ASPECT-7.5
CHASE CONSCIOUSNESS IMPULSES AS CARRIERS FOR LIFE VIRTUES RECAPITULATION

1. It may be recapitulated that intelligence field is created by the transcending mind by transcending through the manifested solid boundary of Creator's space.
2. It may further be recapitulated that consciousness field is created by the transcending mind by transcending through the hyper solid boundary of transcendental worlds.
3. It may further be recapitulated that the consciousness field emerges within Creator's space and the consciousness impulses are superimposed upon the intelligence waves.

PROJECTIONS

4. It would be a blissful exercise for the sadkhas to project further transcendence of the transcending mind through super hyper solid boundaries of self referral domains of 6-space and thereby there being the emergence of the life virtue within consciousness field.
5. It further be a blissful exercise for the sadkhas to

project the superimposition of life virtues upon consciousness impulses parallel to the consciousness impulses getting superimposed upon the intelligence waves.

EXERCISES

1. One shall attempt chase of the transcendence phenomenon happening of life virtues emerging within consciousness field and getting superimposed upon consciousness impulses as transcendental carriers.
2. One shall evaluate one's comprehension and understanding of the transcendence phenomenon happening of life virtues emerging within consciousness field and getting superimposed upon consciousness impulses as transcendental carriers.

■

Transcendental carriers within rays of the Sun

Scritpture preserve the experiences of the senior sadkhas about the existence of the phenomenon of transcendental carriers within rays of the Sun.

Scripture further preserve the ancient wisdom as that the transcendental carriers are embedded with Brahman intelligence.

As such, the sadkhas shall permit their mind to transcend to glimpse this phenomenon.

■

ASPECT-7.6
CHASE LIFE VIRTUE RIDING TRANSCENDENTAL CARRIERS OF RAYS OF THE SUN

1. One shall permit the transcending mind to glimpse the transcendence phenomenon happening of emergence of life virtues within the consciousness field riding consciousness impulses and being carried through rays of the Sun to orb of the Sun.
2. One shall go in trans time and again and to attempt to glimpse and this transcendence phenomenon happening and to continue evaluating one's experiences at every attempt.

■

Transcendence for Enlightenment

Firstly as within human frame,

Secondly as within orb of the Sun (D), and,

Thirdly as the *Ati-Vahkas* path,

of first fold uptil *Sapat-Rishi Lok* and,

then of the final fold,

from *Sapat-Rishi-Lok*, to the *Brahman* domain.

(From book Glimpses of Vedic mathematics)

■

ASPECT-7.7
CHASE UNITY STATE OF CONSCIOUSNESS

1. The unity state of consciousness is the ultimate seventh state of consciousness field.
2. This state of consciousness deserves to be chased as sapt rishi lok / 7-space with 5-space / transcendental worlds in the role of dimension.
3. The transcendental worlds in the role of dimension create a transcendental order for the unity state of consciousness.
4. The transcendental order supplies five folds transcendence ranges for each of the dimensions of the transcendental worlds.
5. The transcendental worlds being of solid dimensional order, as such each of it acquires transcendental features because of the transcendental range.
6. With this the transcendence phenomenon happening takes place within each of the solid dimension of the transcendental worlds and thereby there emerge transcendental skylines of the transcendental worlds.

EXERCISES

1. It would be blissful exercise for the sadkhas to chase the transcending mind creating intelligence field and their emerging perfection of intelligence for the sadkhas.
2. It would further be a blissful exercise to chase the transcending mind attaining consciousness field within intelligence field and sequentially there being a unity state of consciousness for the sadkhas.
3. It would further be a blissful exercise to chase the transcending mind transcending through the transcendental boundary of self referral domains and transforming itself into 'self referral Being' and attaining the emergence of 'LIFE -VIRTUES' within consciousness field.
4. It would further be a blissful exercise to glimpse the transcendental carriers of rays of Sun carrying life virtues firstly to orb of the Sun and from there onwards to the sapt rishi lok / pole star as unity state domain.

■

FEATURE-8
EMERGENCE OF TRANSCENDENTAL WORLDS WITHIN CREATOR'S SPACE

FEATURE-8

Eighth feature shall be the chase of emergence of transcendental worlds (5-space of solid order) within Creator's space (4-space of spatial order).

ASPECT-8.1
UNITY STATE OF CONSCIOUSNESS LEADS TOTRANSCENDENTAL ORDER

INTRODUCTORY

1. Here we have reached the eight feature, as the fourth and last feature of the 'middle pair of fortnights' of the course.
2. The three features of this pair of fortnights of the course just completed a previous day are:

Fifth feature	Reality of 5-space within 4-space
Sixth feature	Fulfilling solid quantifiers within creator's space
Seventh feature	Manifestations of solid quantifiers

CHASE STAGE

3. we have reached a phase and stage of the course where we have to comprehend and understand as to how the unity state of consciousness leads to the transcendental order.
4. There are seven states of the field of consciousness, namely, waking state of consciousness (1), dream state of consciousness

(2), deep sleep state of consciousness (3), transcendental / turia state of consciousness (4), cosmic state of consciousness (5), God state of consciousness (6) and unity state of consciousness (7).

5. Parallel geometric formats of these seven states are of seven transcendental ranges (N-2, N-1, N, N+1, N+2), N=1 to 7.
6. Of these, the last, seventh state of consciousness, namely, unity state of consciousness, is of transcendental order, as that 5-space is in the role of dimensional order.
7. Here, it may be relevant to appreciate as that the consciousness field is emerging within intelligence field.
8. As such, the consciousness state phenomenon deserves to be chased as emergence of 5-space within 4-space as a set of transcendental features of transcendental worlds fulfilled within 4-space with supply of solid quantifiers from origin source reservoir of the transcendental worlds.
9. The solid order fulfilled within Creator's space gets superimposed upon the spatial order.
10. It makes the processing systems as of a pair of dimensional order being simultaneously available.
11. It is like 'interval and square' both being simultaneously available within a 'square'.
12. It is, because of it, that, when the consciousness state are sequentially arranged, the fourth step of this sequencing, gets designated as 'transcendental consciousness state / turia state'.

13. It is a sequential phase and stage when 5-space is expressing itself as transcendental features within 4-space.
14. And with it, the solid order of the transcendental worlds gets superimposed upon a spatial order.
15. And further, the linear order gets carried along the spatial manifestation layer.
16. This transcendental phenomenon of happening of transcendental order of the unity state of consciousness, may be sequentially chased.
17. This sequential chase is to be as 1-space of (-1) space dimensional order as a waking state order.
18. Further sequential chase is to be as 2-space of (0) space dimensional order as a dream state order.
19. 3-space of 1-space as dimensional order of a deep sleep state order.
20. 4-space of 2-space as dimensional order of a turia state.
21. 5-space of 3-space as dimensional order as a cosmic state mundane order.
22. 6-space of 4-space as dimensional order as a God state creative order.
23. 7-space of 5-space as dimensional order as a unity state transcendental order.

EXERCISES

Sadkhas shall permit their transcending mind to glimpse the transcendental order of the unity state of consciousness / 5-space in the role of dimension of 7-space.

Sadkhas shall permit their transcending mind to glimpse the sequential phenomenon of orders of consciousness state beginning with waking state and reaching up till the unity state.

Sadkhas shall also intellectually chase the phenomenon of orders of consciousness states and to validate their experiential glimpsing thereof. ■

Human frame with one head as a pair of eyes is,

(a) the creation, of the creator,

(b) about whose enlightenment, the scriptures are,

(a) full of preservation of the *Shrutis*,

(b) as the experiential expressions,

(c) shared by the *Sadhkas*,

(d) having glimpsed the transcendental world,

as that,

(a) the creator the supreme,

(b) *Lord Brahma* (ठ),

is the,

(i) four head lord,

(ii) with each head equipped with a pair of eyes and, that creator has,

(i) created the creations,

(ii) on His idol's format,

(iii) and of the order and discipline,

(iv) which permits transcendence,

(a) to its transcendental base of transcendental world,

(b) with *Lord Shiv*(ड) as over lord.

(From Book Glimpses of Vedic mathematics)

■

ASPECT-8.2
TRANSCENDENTAL ORDER LEADS TO TRANSCENDENCE WITHIN DOMAIN AS WELL AS WITHIN DIMENSION, TO THEIR SKYLINES

1. One feature of the transcendental order / 5-space in the role of dimension, is that it makes available a transcendental range of 5 steps long measuring rod along each of the transcendental dimension of unity state of consciousness.
2. This phenomenon of transcendental range being available along each of the seven transcendental dimensions is of transcendental features and character as that within each of the transcendental dimension of the unity state, there takes place a transcendence phenomenon which takes to dimension of dimension level, designated as skylines of the transcendental worlds.
3. This happening of transcendence phenomenon within a transcendental dimension deserves to be chased along each of the five solid dimensions of the transcendental dimension role being played by 5-space.
4. Each of the five dimensions of a transcendental dimension being of a solid order, so the transcendence phenomenon within a

transcendental dimension firstly takes to the solid dimensions and then to the linear dimensions of the solid dimension.

5. This happening makes transcendence phenomenon of the order of artifice 5 x 3 x 1 =15 for taking transcendence to linear order as skyline of the transcendental worlds within solid dimensions of the transcendental dimension itself.
6. As such this phenomenon, ultimately carries manifested values along the intelligence waves into sapt rishi lok / 7-space with transcendental dimensional order creating consciousness field with its impulses as transcendental carriers.
7. The transcendental carriers carry the life virtues of self referral domains of 6-space.
8. This phenomenon of transcendental carriers is the phenomenon of the transcending mind even transcending through the consciousness field and attaining self referral domains of 6-space.
9. This, is the phase and stage of transcendental carriers within rays of the sun / 6-space domain.
10. This attainment is designated as the attainment of the skylines of the transcendental worlds supplying transcendental ranges for transcendental features to superimpose consciousness impulses upon the intelligence waves of the Creator's space, and also because of the spatial order of Creator's space, making available of being carried with reversal of orientation resulting into the consciousness impulses carrying the intelligence waves as well.
11. It is this feature of the happening of transcendence

phenomenon within transcendental dimensions, may also be chased as that the solid dimensional order of transcendental dimensions is availing Creator's space as its sustainer.

12. This way, the emergence of the transcendental worlds within Creator's space and the emergence of Creator's spaces at origin sources of the solid dimensional order, makes the Creator's space being of unique transcendental features.

EXERCISES

1. The sadkhas shall permit their transcending mind to glimpse the emergence of transcendental worlds within Creator's space.
2. Further the sadkhas shall permit their transcending mind to glimpse the emergence of Creator's space as origin base source of solid dimensional order.
3. Sadkhas shall permit their transcending mind to simultaneously glimpse the emergence of transcendental worlds within Creator's space and the emergence of Creator's space as origin base source of solid dimensional order.
4. Still further the sadkhas shall chase unity state of Creator's space leading to transcendental order leading to transcendence within the transcendental dimension as well as within its dimension of dimension as transcendental skylines.
5. Sadkhas shall validate their experiential chase in the light of the intellectual chase of the dimensional order of 7-space up till its skylines at its dimension of dimension level. ■

ASPECT-8.3
CHASE SKYLINES OF UNITY STATE ALONG ITS FIRST SOLID DIMENSION OF TRANSCENDENTAL DIMENSIONAL ORDER

1. It would be a blissful exercise for the sadkhas to chase intellectually as well as to experience by glimpsing through the transcendental worlds, in their roles as dimensional order of unity state of consciousness.
2. As the transcendental worlds are of solid dimensional order and the transcendental features express themselves within a dimensional frame of 5 solid dimensions, as such the happening of transcendence phenomenon deserves to be sequentially chased from first to the fifth solid dimension of the dimensional frame.
3. The transcendence to dimension of dimension level within transcendental worlds leads to the skylines of the transcendental worlds.
4. As such the transcendence through any of the solid dimension of the transcendental worlds is to lead to skyline of linear format.
5. However, when the dimensions of the dimensional frame are sequenced, and these get designated as first, second, third, fourth and fifth dimension,

these accordingly amounts to be of additional features as being first, second, third, fourth or fifth.

6. The first solid dimensional order acquires linear format for its skyline but the second solid dimensional order, being second, so presumes the existence of the first, and as such it acquires additional features which make the format for the skylines being as of 'spatial order'.
7. This, this way, while for the first dimension, makes the skylines, as to be of linear format but at the phase and stage of second dimension it acquires transition and transformation of spatial format for the skylines of the transcendental worlds.
8. Likewise at the phases and stages of third, fourth, fifth and sixth dimension, there emerge transitions and transformations, as of solid, hyper solid and super hyper solid formats for the skylines of the transcendental worlds.
9. As such the chase of the skylines of the transcendental worlds is to be of sequential features beginning with the linear format and reaching up till the super hyper solid format for establishing the solid dimensional order for the transcendental dimensions of the unity state of consciousness field being chased.
10. This sequential chase, deserves to be sequentially chased and the experiential glimpses of each sequential step deserve to be intellectually validated.

EXERCISES

1. Chase format of skylines of transcendental worlds

while transcending through any of the solid dimension of the transcendental worlds.

2. Chase linear format of skylines as a linear sequential flow of the transcendental features at level of the skylines of the transcendental worlds.
3. Chase spatial format of skylines as a spatial flow of the transcendental features at the level of the skylines of the transcendental worlds availing a pair of linear formats for the skylines supplied by a pair of axes of the transcendental worlds.
4. Chase solid format of skylines as a solid flow of the transcendental features at the level of the skylines of the transcendental worlds availing triple linear formats for the skylines supplied by triple axes of the transcendental worlds.
5. Chase hyper solid format of skylines as a hyper solid flow of the transcendental features at the level of the skylines of the transcendental worlds availing quadruple linear formats for the skylines supplied by quadruple axes of the transcendental worlds.
6. Chase super hyper solid format of skylines as a super hyper solid flow of the transcendental features at the level of the skylines of the transcendental worlds availing all the five quadruple linear formats for the skylines supplied by all the five axes of the transcendental worlds.
7. It shall be a blissful exercise for the sadkhas to permit their transcending mind to glimpse the establishment of solid dimensional order.

■

ASPECT-8.4
CHASE SKYLINES OF UNITY STATE ALONG ITS SECOND SOLID DIMENSION OF TRANSCENDENTAL DIMENSIONAL ORDER

EXERCISES

1. One shall firstly chase transcendence to skylines of the transcendental worlds within any pair of axes without prejudice to either of them being the first or the second.
2. Then, transcendence be chased again through the given pair of axes of the transcendental worlds by having them sequenced as first and second.

In this light, the phenomenon of transcendence through pair of axes leading to linear format for the skylines of transcendence through first axes and spatial format for the skylines of the second axes be chased and the chase experience be intellectually validated and the same be expressed for its evaluation.

■

Skylines

Skylines are dimensions of dimensions.

■

ASPECT-8.5
CHASE SKYLINES OF UNITY STATE ALONG ITS THIRD SOLID DIMENSION OF TRANSCENDENTAL DIMENSIONAL ORDER

EXERCISES

1. One shall firstly chase transcendence to skylines of the transcendental worlds within any three of axes without prejudice to either of them being the first, second or third.
2. Then, transcendence be chased again through the given three axes of the transcendental worlds by having them sequenced as first, second and third.

In this light, the phenomenon of transcendence through three of axes leading to linear formats for the skylines of transcendence through first axis, spatial format for the skylines of the second axis and solid format for the skylines of the third axis be chased and the chase experience be intellectually validated and the same be expressed for its evaluation. ■

Skyline of unity state

Skyline of unity state is solid order which firstly leads to transcendental worlds and then to the unity state. ■

ASPECT-8.6
CHASE SKYLINES OF UNITY STATE ALONG ITS FOURTH SOLID DIMENSION OF TRANSCENDENTAL DIMENSIONAL ORDER

EXERCISES

1. One shall firstly chase transcendence to skylines of the transcendental worlds within any four of axes without prejudice to either of them being the first, second, third or fourth.
2. Then, transcendence be chased again through the given four of axes of the transcendental worlds by having them sequenced as first, second, third and fourth.

In this light, the phenomenon of transcendence through four of axes leading to linear formats for the skylines of transcendence through first axis, spatial format for the skylines of the second axis, solid format for the skylines of the third axis and hyper solid format for the skylines of the fourth axis be chased and the chase experience be intellectually validated and the same be expressed for its evaluation.

■

ASPECT-8.7
CHASE SKYLINES OF UNITY STATE ALONG ITS FIFTH SOLID DIMENSION OF TRANSCENDENTAL DIMENSIONAL ORDER

EXERCISES

1. One shall firstly chase transcendence to skylines of the transcendental worlds within five axes without prejudice to either of them being the first, second, third, fourth or fifth.
2. Then, transcendence be chased again through all the five axes of the transcendental worlds by having them sequenced as first, second, third, fourth and fifth.

In this light, the phenomenon of transcendence through five axes leading to linear formats for the skylines of transcendence through first axis, spatial format for the skylines of the second axis, solid format for the skylines of the third axis, hyper solid format for the skylines of the fourth axis and super hyper solid format for the skylines of the fifth axis be chased and the chase experience be intellectually validated and the same be expressed for its evaluation as to the comprehension of the phenomenon of establishment of solid order for the transcendental worlds.

LINKING NOTE

First pair of fortnights Features 1 to 4	Middle pair of fortnights Features 5 to 8	Last pair of fortnights Features 9 to 12
		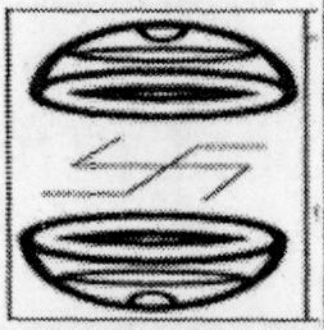
Sathapatya Foundation	Transcendental worlds	Pair of hemispheres

LAST PAIR OF FORTNIGHTS

Features
Fifth fortnight
1. Feature 09 'About synthesis of manifestation layers'
2. Feature 10 'About partitioning of universe'
Sixth fortnight
3. Feature 11 'About distinct universes'
4. Feature 12 'About self referral domain'

INTRODUCTORY

FIRST PAIR OF FORTNIGHTS

First pair of fortnights features focus upon 'FOUR SPACE REALITY' with concentration upon hyper cube 4, and of it, the feature of utmost interest being its origin fold / transcendental core of origin of 4-space as a seat of compactification of whole range of origins.

This takes us to the Sathapatya foundation. Its symbolic expression may be as under, a set up of Shivligam / transcendental stem / pole installed within Creator's space at its transcendental core.

MIDDLE PAIR OF FORTNIGHTS

This pair of fortnights features focus upon the phenomenon of transcendental worlds within Creator's space. The emergence of field within fields is the distinct features of this set up. It is to be comprehended and chased as the phenomenon of transformation of intelligence field into consciousness field. It may be symbolically depicted as follows as transcendental worlds / hyper cube 5 / 5-space / consciousness field.

LAST PAIR OF FORTNIGHTS

This final pair of fortnights features focus upon split of universe around skyline of transcendental worlds into a pair of transcendental worlds within Creator's space sustained by self referral domain (6-space) core of the transcendental worlds.

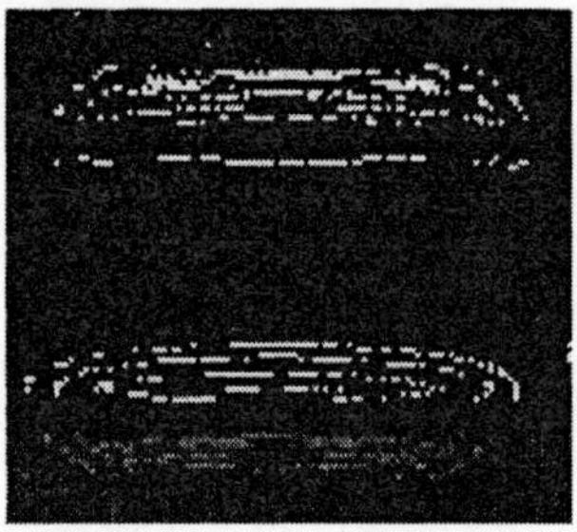

■

FEATURE-9
SYNTHESIS OF MANIFESTATION LAYERS

LINKING NOTE

"**Ninth feature** shall be a chase of synthesis of manifestation layers with transcendental base. This feature also shall be chasing the transcendental phenomenon of the origin fold acquiring new rule of dimension fold of next manifestation layer because of the transcendental base beneath the origin fold."

THIS FEATURE MAY BE PHASED FOR ITS CHASE AS FOLLOWS

1. Synthesis of manifestation layers along skylines of transcendental worlds.
2. Spatial axes format for the skylines of the transcendental worlds.
3. Ascendance through inner folds of the transcendental worlds.
4. Ascendance along both axes of the spatial order.
5. Ascendance is to be chased for its feature being reversal of transcendence.
6. Cosmic phenomenon of 'TRANS' sustained by transcendental skylines.
7. Coordination of spatial worlds by the transcendental core of Creator's space.

ASPECT-9.1
SYNTHESIS ALONG SKYLINES

1. This phase and stage of the course takes us to the stage of chasing synthesis of manifestation layers along the skylines of the transcendental worlds within Creator's space.
2. Manifestation layers avail four fold manifestation format with each fold / quarter of the format availing full fledged four fold format because of the availability of the spatial order of Creator's space.
3. Chase of this set up of manifestation layers availing four fold manifestation format with each fold further availing full fledged four fold format within Creator's space being of a spatial order shall be leading us to the format features as of following expression:

   N-2, N-1, N+0, N+1

   N-1, N+0, N+1, N+2

   N+0, N+1, N+2, N+3

   N+1, N+2, N+3, N+4
4. This provides seven steps long range

(N-2, N-1, N+0, N+1, N+2, N+3, N+4)

for synthesis of a pair of manifestation layers (N-2, N-1, N+0, N+1) and (N+1, N+2, N+3, N+4).

5. This synthesis range is focusing upon the origin fold of first manifestation layer transiting and transforming its role as of dimension fold of the second manifestation layer.
6. As such, it be comprehended well as that the synthesis is a transcendence phenomenon happening at the origin fold (N+1)-space, and it again attaining the role of origin fold at three steps ahead, that is, at (N+4)-space.
7. This three steps long jump is provided by the solid order of the transcendental worlds by transcendence phenomenon happening of transition and transformation from one solid dimension to the next solid dimension.
8. In the context, it also would be relevant to focus attention for comprehension of the spatial set up of four fold manifestation format with each fold itself as well being a full fledged four fold manifestation. This 4 x 4 set up, as a square, with sequential progressions of four folds of four consecutive manifestation layers shall be transforming the square set up into a solid / cubic set up as that.

1, 1, 1, 1	1, 2, 3, 4
1, 1, 1, 1= 4 × 4, and	2, 3, 4, 5 = 4 × 4 × 4
1, 1, 1, 1	3, 4, 5, 6
1, 1, 1, 1	4, 5, 6, 7

9. It would be a blissful chase as that the set up of

above format of four folds with each fold as well being of four folds manifests a format for simultaneous manifestation of four consecutive manifestation layers, as follows.

-1, 0, 1, 2

0, 1, 2, 3

1, 2, 3, 4

2, 3, 4, 5

10. It would further be blissful to chase above set up by focusing upon the spatial flow (2-space) along the diagonal, which means that the affine set up of points as a side of a square transforms itself as affine spatial set up along the diagonal, as that, 0-space as dimension of a format of a side of the square, is structuring out 2-space as domain of a format of a diagonal of the square.

11. It would further be blissful to chase the above set up as a split up of a square into a pair of tri angles on either side of the diagonal, as a skyline of the transcendental worlds of solid dimensional order which at its dimension of dimension being of a linear set up.

12. It would further be a blissful to chase as that the diagonal can be approached from either side as of sequential steps, as follows:

From one side		From second side	
a.	–1	(1)	5
b.	0, 0	(2)	4, 4
c.	1, 1, 1	(3)	3, 3, 3

13. The skylines of Creator's space at its dimension of

dimension level lead to '0-space' which levels all entities as $(A^0)^N = 1$, for all A and N.

14. This, as such, leads to a measuring rod of units, viz,

    $(A^0)^1 + (A^0)^2 + (A^0)^3 + (A^0)^4 + —$

15. This, along the diagonal of four consecutive manifestation layers with 'different roles for 0-space', shall be manifesting the affine measuring rod for the Creator's space at its skylines as follows:

    -3, -2, -1, 0

    -2, -1, 0, 1

    -1, 0, 1, 2

    0, 1, 2, 3

16. It would be a blissful exercise to chase the above set up on either side of the affine 0-space values diagonal as four fold measuring rod of uniform values for all the four quarter as a pair of tri-angles of opposite orientations being sustained by the diagonal as skyline of Creator's space.

17. It would further be a blissful exercise to chase a pair of tri-angles on both sides of the diagonals as of opposite orientations and as replicas of 3-space set ups of pair of 3 dimensional frames of half dimensions into which a three dimensional frame would split.

18. It would further be a blissful exercise to chase the above pair of set ups within Creator's space of spatial order approaching spatial order as artifice two as 'one', and thereby the universe of transcendental worlds structured by solid order

leading to a pair of universes of transcendental worlds around the transcendental skyline.

EXERCISES

One shall intellectually validate one's experiential glimpsing of the transcendence phenomenon happening of split of a universe of transcendental worlds into a pair of transcendental worlds of the format of seven steps long synthetic range for synthesis of manifestation layers with origin fold of first manifestation layer transiting and transforming its role from origin fold to that of dimension fold of the second manifestation layer.

■

Ways to approach the Discipline:

I. On the artifice of 5

II. On tri-Monad format

III. While in-trans glimpsing the transcendental world (卐)

IV. Worshiping the Idol of ***Lord Shiv*** (卐)

V. Working out transcendental phenomena on the format of hyper cube 5

VI. With knowledge of *Lord Shiva*'s (卐) family

VII. Chasing the transcendence traditions of *Lord Brahma* (B), creator the supreme

VIII. Sharing ambrosia of bliss with the *Sadhakas* having glimpsed the transcendental worlds (卐).

IX. Chiseling third eye by glimpsing the transcendental world (卐) by permitting the mind to transcend and also by intellectually chasing the artifice format of *Sathaptya Upved, Mansara,* which begins with the format of idol of *Lord Brahma*(B) and attains chiseling of third eye.

X. Privilege of chasing the *Divya Ganga* flow

XI. Grace of Lord of transcendental worlds (卐)

(From Book Glimpses of Vedic mathematics) ■

ASPECT-9.2
SKYLINES ALONG SPATIAL AXES

1. This phase and stage of the course takes us to the stage of chasing synthesis of manifestation layers as skylines format of spatial axes.
2. Along the spatial axes, a three dimensional frame splits into a pair of three dimensional frames of half dimensions.
3. Spatial order as artifice '2' being the unit shall be splitting three dimensional frame into a pair of three dimensional frames of full units of artifice '1'.
4. With this, each of the three dimensional frames of above pair shall be manifesting a solid order for structuring a transcendental world.
5. This, this way, shall be leading us to transcendence phenomenon happening of a universe of transcendental worlds transiting and transforming as a pair of universes of transcendental worlds within Creator's space because of its spatial order of pair of axes each of which as a transcendence range leading to a transcendental world.

6. It would be a blissful exercise to chase the transcendence phenomenon happening of transcendence range along each of the axis of spatial order of Creator's space leading to a transcendental world.
7. It would further be a blissful exercise to chase the transcendence phenomenon happening of a transcendental world resulting into a pair of transcendental worlds sustained as a transcendence phenomenon happening along the transcendence ranges availed by the pair of axes of the spatial order of the Creator's space.
8. It would further be a blissful exercise to glimpse the transcendence phenomenon happening of a transcendental worlds universe resulting into a pair of universes, which further resulting into a double pair of universes and the sequential progression to continue without any blocking within Creator's space.
9. It would further be a blissful exercise to glimpse the transcendence phenomenon happening of the transcendental worlds universe resulting into a pair of universes because of the Creator's space of spatial order being the origin source reservoir base of the solid order of the transcendental worlds.
10. It would further be a blissful exercise to sequentially chase, starting with a Creator's space, as step 1, it being fulfilled with solid quantifiers from its source origin as step 2, the solid quantifiers manifesting as solid manifestation layer within Creator's space as step 3, the Creator's space as origin fold of solid manifestation layer

fulfilling spatial order within linear order solid domain as step 4, and the solid dimensional frame splitting into a pair of solid dimensional frame as step 5 and finally each of the solid dimensional frame structuring transcendental world within Creator's space as step 6.

EXERCISES

One shall intellectually validate one's experiential glimpsing of the above transcendence phenomenon happening of split of a universe of transcendental worlds into a pair of transcendental worlds in sequential steps: starting with a Creator's space, as step 1, it being fulfilled with solid quantifiers from its source origin as step 2, the solid quantifiers manifesting as solid manifestation layer within Creator's space as step 3, the Creator's space as origin fold of solid manifestation layer fulfilling spatial order within linear order solid domain as step 4, and the solid dimensional frame splitting into a pair of solid dimensional frame as step 5 and finally each of the solid dimensional frame structuring transcendental world within Creator's space as step 6.

■

Heavens and skies

It is blissful to know that

NVF (HEAVEN)=NVF (SKY)

■

ASPECT-9.3
ASCENDANCE THROUGH INNER FOLDS OF THE TRANSCENDENTAL WORLDS

RECAPITULATION

1. Skylines manifest as dimension of dimension of the universe.
2. The skylines of transcendental worlds are of linear order as 1-space, as dimension leads to 3-space which structures 5-space / transcendental worlds.
3. 1-space accepts manifestation format for being a four fold manifestation layer (-1, 0, 1, 2).
4. The coordination of (-1) space and (+1) space as dimension and domain results into reversal of transcendence at skyline and continuation of the transcendence phenomenon happening continuing as 'ascendance phenomenon happening' through the inner layers of the transcendental worlds.
5. The skylines of Creator's space (4-space) are of affine (0-space) being dimension of dimension of 4-space.
6. The reversal of transcendence phenomenon happening into ascendance phenomenon happening within Creator's space takes back from dimension of dimension state of 0-space to

structured spatial order / 2-space as domain structured by 0-space during ascendance phenomenon happening at the skyline.

7. It would be a blissful exercise to chase the transcendence phenomenon happening transforming into ascendance phenomenon happening within Creator's space and there being ascendance through the inner folds of the transcendental worlds emerging within Creator's space.

EXERCISES

One shall intellectually validate one's experiential glimpsing of the above transcendence phenomenon happening transforming into ascendance phenomenon happening within Creator's space and there being ascendance through the inner folds of the transcendental worlds emerging within Creator's space.

Transcendental format of mind of Sadhaka.

1. The transcendental format of the mind of the *Sadhaka* being fortunate and privileged having glimpsed the transcendental worlds is the format of five heads each equipped with three eyes.
2. This is the format of the mind with setting of *Vishav-Rupa* transcending through all the five solid dimensions of the format of the transcendental worlds and ascending with ambrosia of bliss of the transcendental world through all the five dimensions of the tran scendental world.
3. This is the transcendence phenomena along all the five solid dimensions of the transcendental world as 15 *Shalokas* range is enlightenment of *Sri Saraswati Stotram*.

(From Book Glimpses of Vedic mathematics) ■

ASPECT-9.4
ASCENDANCE ALONG BOTH THE AXES OF THE SPATIAL ORDER

1. One shall permit one's transcending mind to glimpse the transcendence phenomenon happening along both the axes of the spatial order as independent happening as transcendence ranges of each of the axis is to lead to independent transcendental world.
2. One shall further permit one's transcending mind to glimpse the transcendence phenomenon happening along first axis of the spatial order transiting and transforming by reversal of orientation into ascendance phenomenon happening at the skylines of the transcendental world achieved along the transcendence range of the axis.
3. One shall further permit one's transcending mind to glimpse the transcendence phenomenon happening along second axis, after glimpsing along the first axis, and to experience the transcendence phenomenon happening transiting and transforming along second axis as well by reversal of orientation into ascendance phenomenon

happening at the skylines of the transcendental world achieved along the transcendence range of the second axis.

4. One shall further permit one's transcending mind to simultaneously glimpse the transcendence phenomenon happening along both the axis and to experience the transcendence phenomenon happening transiting and transforming along both axes by reversal of orientations into ascendance phenomenon happenings at the skylines of the transcendental worlds achieved along the transcendence ranges of the both axes.
5. One shall further permit one's transcending mind to simultaneously glimpse the transcendence phenomenon happening within a pair of independent transcendental worlds sustained within the Creator's space by the spatial order of the Creator's space.

EXERCISES

One shall intellectually validate the experience of glimpsing the transcendence phenomenon happenings of reversal into ascendance phenomenon happenings at skylines of a pair of independent transcendental worlds within Creator's space sustained by the spatial order of the Creator's space.

■

Pair of transcendental worlds

Wonders of spatial order of creator's space may be glimpse as manifestation of distinct transcendental worlds within northern and sotheren hemispehre.

■

ASPECT-9.5
ASCENDANCE IS TO BE LOOKED FOR REVERSAL OF TRANSCENDENCE

1. One shall permit one's transcending mind to chase the ascendance phenomenon happening at skylines of the transcendental worlds as reversal of the transcendence phenomenon happening through the manifested hyper solid boundaries of the transcendental worlds.
2. One shall also permit one's transcending mind to chase transcendence phenomenon happening at manifested boundaries of the transcendental worlds as reversal of the ascendance phenomenon happening through the inner layers of the transcendental worlds.
3. One shall permit one's transcending mind to chase the transcendence and ascendance phenomenon happenings within transcendental worlds as a continuous process as a pair of halves of the circumference of a circle enveloping the spatial order of Creator's space.
4. One shall permit one's transcending mind to chase the transcendental phenomenon of sustenance of pair of transcendental worlds by the spatial order of the Creator's space.

5. One shall permit one's transcending mind to chase the transcendental phenomenon of the pair of transcendental worlds transiting and transforming as a single transcendental world like the pair of halves of the circumference constituting a single whole circumference of continuity set up.

EXERCISES

1. One shall intellectually validate experiential chase of the transcendental mind glimpsing ascendance phenomenon happening at skylines of the transcendental worlds as reversal of the transcendence phenomenon happening through the manifested hyper solid boundaries of the transcendental worlds.
2. One shall intellectually validate experiential chase of the transcendental mind glimpsing the transcendence phenomenon happening at manifested boundaries of the transcendental worlds as reversal of the ascendance phenomenon happening through the inner layers of the transcendental worlds.
3. One shall intellectually validate experiential chase of the transcendental mind glimpsing the transcendence and ascendance phenomenon happenings within transcendental worlds as a continuous process as a pair of halves of the circumference of a circle enveloping the spatial order of Creator's space.
4. One shall intellectually validate experiential chase of the transcendental mind glimpsing the transcendental phenomenon of sustenance of pair

of transcendental worlds by the spatial order of the Creator's space.

5. One shall intellectually validate experiential chase of the transcendental mind glimpsing the transcendental phenomenon of the pair of transcendental worlds transiting and transforming as a single transcendental world like the pair of halves of the circumference constituting a single whole circumference of continuity set up.

Triloky (⊞) and Trinity of Gods (⌗, §, ⋈).

Sadhakas having an intensified urge to glimpse the disciplines of *Vedic* Mathematics science and technology through transcendental glimpse having protections of the transcendental reservoir with a transcendental seat at the core the transcendental reservoir shall transcend time and again with manifested format of *Vishav-Rupa* of our *Triloky* (⊞) to its transcendental base by glimpsing the transcendental bases of *Trinity* of Gods (⌗ § ⋈).

Privileged and fortunate would be the *Sadhakas* Transcending and glimpsing the happening of the transcendental phenomena of *Vishav-Rupa* of three heads, each with a pair of eyes, transcending and glimpsing the creators way and going transcendental, chiseling third eye, having transcendental state of consciousness, perfecting intelligence and glimpsing *Brahm* as *Nav-Brahm* within five folds at the inner most seat of the transcendental world and there by the manifested format of his with pair of eyes going transcendental as a head with three eyes and with it *Triloky* (⊞) going transcendental as *Trinity* of Gods(⌗ § ⋈) with *Sapat Rishi Lok* at *Brahman Ati-Vahkas* carrying Being to the *Brahman* domain.

(From Book Glimpses of Vedic mathematics)

ASPECT-9.6
COSMIC PHENOMENON OF TRANS SUSTAINED BY TRANSCENDENTAL SKYLINES

1. One shall permit one's transcending mind to glimpse the cosmic phenomenon of trans of transcending mind glimpsing the transcendental universe transforming itself as a pair of transcendental universes and also the pair of transcendental universes unifying as a single transcendental universe.
2. One shall go in trans time and again and to be blissful of the ambrosia of bliss of this cosmic phenomenon.

EXERCISES

1. One shall intellectually validate the experiences of glimpsing the cosmic phenomenon of trans of transcending mind glimpsing the transcendental universe transforming itself as a pair of transcendental universes and also the pair of transcendental universes unifying as a single transcendental universe.
2. One shall express and evaluate one's experiences and their intellectual validations and to share the same with others in similar states. ■

ASPECT-9.7
COORDINATION OF SPATIAL WORDS BY THE TRANSCENDENTAL CORE OF CREATOR'S SPACE

Final Exercises of the feature

1. It would be a best exercise to chase the way spatial order is sustained by the transcendental core of the Creator's space by fulfilling Creator's space domain with solid quantifiers.
2. One shall attempt this chase time and again and to evaluate the same for self-certification of comprehension and understanding of the feature of synthesis of manifestation layers with emergence of skylines.

■

Think about

Geometries of 6-space

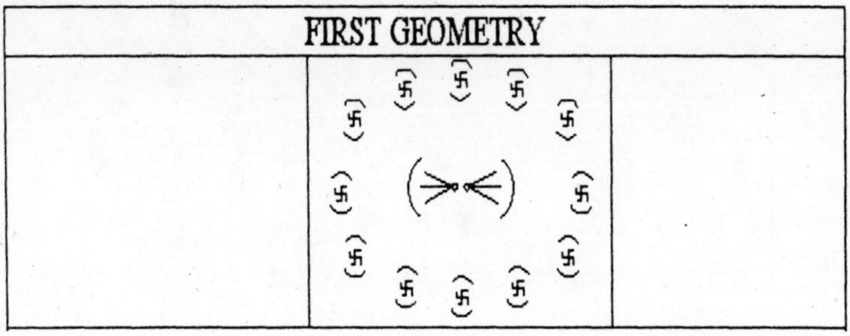

Geometries 2nd to 13th

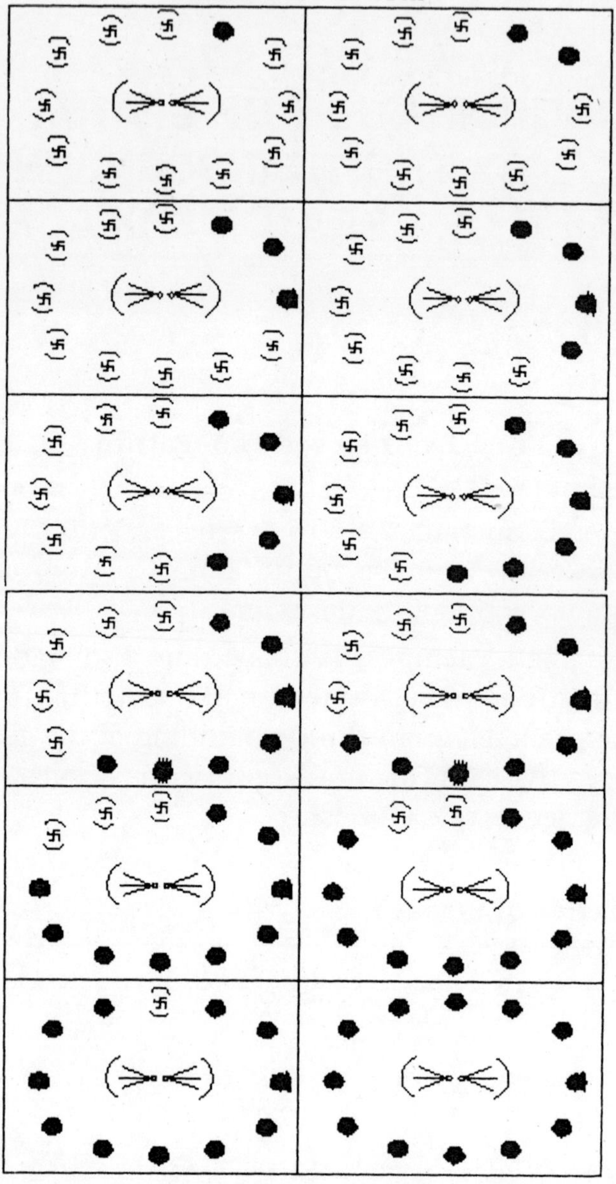

FEATURE-10
PARTITIONING OF UNIVERSE AROUND TRANSCENDENTAL SKIES

First pair of fortnights Features 1 to 4	Middle pair of fortnights Features 5 to 8	Last pair of fortnights Features 9 to 12

Sathaparva Foundation	Transcendental worlds	Pair of hemispheres

FEATURE-10

PARTITIONING OF UNIVERSE AROUND TRANSCENDENTAL SKIES

Tenth feature is about the transition and transformation from transcendental worlds to transcendental skies manifesting at dimension of dimension level, around which universe partitions into a pair of universes.

ASPECT-10.1
INTRODUCTORY

1. The origin source reservoir of Creator's space supplies transcendental features of solid dimensional order into Creator's space.
2. This supply of transcendental features of solid dimensional order is there as fulfilling of Creator's space domain with solid quantifiers.
3. The solid quantifiers manifest within Creator's space a solid manifest layer.
4. The spatial order of Creator's space splits a three dimensional frame (of solid manifestation layer) into a pair of three dimensional frames of half axes.
5. The spatial order accepts artifice '2' as a unit, and as such the pair of 3 dimensional frame of half dimensions / axes emerges to be of 'artifice 1' as dimensional units for both three dimensional frames.
6. The three dimensional frame, essentially being of a linear order / '1-space / artifice 1, playing the role of dimension, and here within Creator's space,

the spatial order supplying 'artifice 1' units even for 'half axis', as such the pair of three dimensional frames of such 'half axes' emerge to be a three dimensional frame of 'full linear axes', and therefore, each of the three dimensional frame, as solid order shall be structuring an independent transcendental world / 5-space domain.

PAIR OF TRANSCENDENTAL WORLDS

7. This feature of creation of a pair of transcendental worlds within the Creator's space of spatial order, deserves to be chased, as a phenomenon of transcendental features emerging within Creator's space domain.
8. This phenomenon of transcendental features emerging within Creator's space domain as a pair of transcendental worlds deserves to be chased as the phenomenon happening at skylines of the transcendental worlds within Creator's space.
9. The split of three dimensional frame into a pair of three dimensional frames of half dimensions deserves to be chased as emergence of a pair of hemispheres within Creator's space.
10. The Creator's space being of a spatial order, and the half dimensions of three dimensional frames being of 'artifice 1 unit', as such, each of the three dimensional frame of such units, and as a consequence thereof, each of the hemisphere sustains a complete three space set up, which as a solid order would be structuring a transcendental world.

EXERCISES

1. One shall chase the split of a three dimensional frame into a pair of three dimensional frames of half dimensions.
2. One shall further chase the spatial order of Creator's space supplying 'artifice 1 units' for half dimensions, and thereby each of the three dimensional frames of half dimensions emerging to be a complete three dimensional frame of full linear dimensional order.
3. One shall further chase the phenomenon of emergence of a pair of hemispheres within Creator's space.
4. One shall further chase the phenomenon of emergence of a complete three dimensional set up of linear dimensional order within each of the hemispheres.
5. One shall further chase the phenomenon of structuring of transcendental worlds by the solid order of the hemisphere.
6. One shall chase the phenomenon of emergence of a pair of transcendental worlds within Creator's space.

■

Know more and more about the sun

Sun is the atman / soul / Vishnu lok / self referral domain / core of transcendental worlds / 6-space / hyper cube-6.

■

ASPECT-10.2
FULL TRANSCENDENTAL WORLD WITHIN FIRST HEMISPHERE

INTRODUCTORY

1. The hemispheres emerging within Creator's space of sustained by linear dimensional order of 'artifice 1 units' supplied by the spatial order of the Creator's space.
2. The linear dimensional order of hemispheres structures them a 'solid dimensional order'.
3. The emerging solid dimensional order further structures a transcendental world / 5-space.

FIRST HEMISPHERE

4. The first hemisphere, as such shall be structuring a first transcendental world.
5. The transcendental world structured by the solid order of the first hemisphere shall be a full fledged transcendental world as the linear order of the hemisphere is availing 'artifice 1' units supplied by spatial order as half of two.
6. The transcendental world structured by first hemisphere is a full fledged transcendental world

as independent transcendental world within a Creator's space.

EXERCISES

1. Chase first hemisphere as of 'artifice 1' units linear dimensional order.
2. Chase first hemisphere as a solid dimensional order.
3. Chase first hemisphere as solid dimensional order structuring a transcendental world / 5-space.
4. Chase transcendental world structured by first hemisphere as a full fledged transcendental world of independent existence within Creator's space.

■

Chase mental states

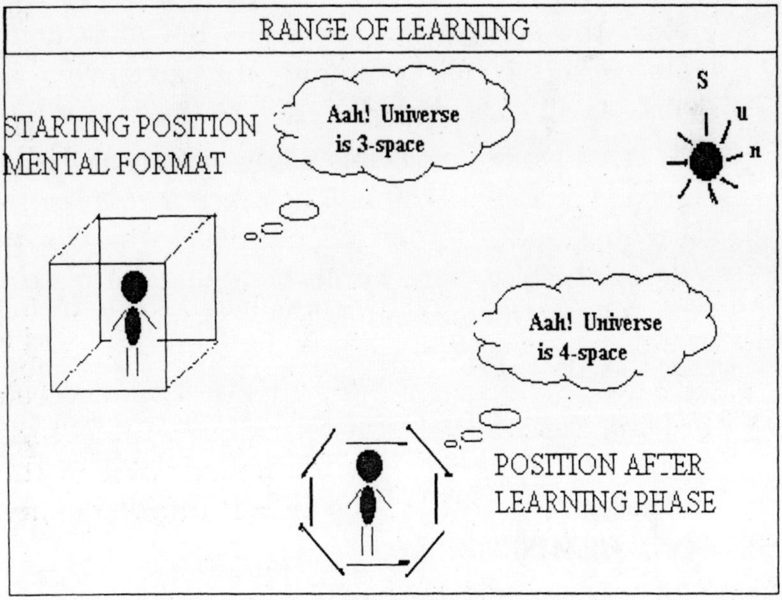

■

ASPECT 10.3
FULL TRANSCENDENTAL WORLD WITHIN SECOND HEMISPHERE

INTRODUCTORY

1. Both the axes of the spatial order are independent axes, and either of it may become the first axis.
2. Ones the particular axis is accepted as the first axis, the second automatically becomes of the status of 'second', presuming the existence of the 'first axis'.
3. As such while processing along or in terms of the second axis, firstly it may be taken as of features, set ups and attainments parallel to that of the first axis, but there would be additional features for it being the second axis.

SECOND AXIS

4. The second axis presumes the existence of first axis, and as such, the second hemisphere is to presume the existence of first hemisphere.

SECOND HEMISPHERE

5. The second hemisphere presumes the existence of the first hemisphere.

6. As such the emergence of second transcendental world within second hemisphere is to be presumed as second independent emergence of the transcendental world, the first having already emerged independently within first hemisphere.

EACH TRANSCENDENTAL WORLD BEING FULL FLEDGED UNIVERSE

7. Both the transcendental worlds of both hemispheres are independent full fledged transcendental worlds as both hemispheres are full fledged solid dimensional orders because of the complete linear dimensional order formats available for both the hemispheres.
8. It would be a blissful experience to permit the transcending mind to glimpse the phenomenon of a pair of full fledged transcendental worlds emerging and existing independently within Creator's space.

EXERCISES

Chase the phenomenon of a pair of full-fledged transcendental worlds emerging and existing independently within Creator's space.

Glimpse and be blissful

■

ASPECT 10.4
SUSTENANCE OF HEMISPHERES BY SPATIAL ORDER

1. The sustenance of hemispheres by the spatial order of the Creator's space deserves to be chased from the phase and stage of emergence of solid quantifiers from within the origin source reservoir of the Creator's space.
2. The solid quantifiers fulfilled within Creator's space domain avail the manifestation format supplied by the spatial order of the Creator's space.
3. The solid quantifiers manifest as solid manifestation layer, as a solid domain with hyper solid base, illustratively, the cube, a set up within a three dimensional frame of linear dimensional order.
4. The spatial order, as a set up of a pair of dimensions, splits a three dimensional frame into a pair of three dimensional frame and that to of 'artifice-1' units.
5. It is this feature of the spatial order of Creator's space supplying 'artifice-1' units for the 'half dimensions' of both three dimensional frames into which a three dimensional frame splits within a

Creator's space, deserves to be chased and to be glimpsed and comprehended well.

6. It is a transcendental phenomenon of sustenance of a sphere as a pair of hemispheres by the spatial order of the Creator's space.
7. It is a transcendental phenomenon of a transcendental universe / 5-space splitting into a pair of transcendental universes with splitting off the solid dimensional order itself.
8. It is a transcendental phenomenon of a pair of transcendental universes unifying as a single transcendental universe with unification of pair of solid dimensional orders into a unified solid dimensional order.
9. It is a transcendental phenomenon of spatial order of Creator's space sustaining the solid order of the transcendental worlds as a pair of solid orders attained by the pair of linear axes of the spatial order of the Creator's space.
10. It is a transcendental phenomenon of transcendental worlds emerging within Creator's space and the Creator's space emerging with the split of the dimensional order of the transcendental worlds.
11. It is a transcendental phenomenon of Creator's space sustaining the transcendental worlds and the transcendental worlds sustaining the Creator's space.
12. It is a transcendental phenomenon of solid order being sustained by the spatial order and the spatial order is being sustained by the solid order.

13. It is a transcendental phenomenon of solid manifestation layer being sustained by the spatial manifestation layer and the spatial manifestation layer sustaining the solid manifestation layer.
14. It is a transcendental phenomenon of transcendence range sustaining the manifestation format and the manifestation format sustaining the transcendence range.
15. It is a transcendental phenomenon of transcendence being sustained by the ascendance and the ascendance being sustained by the transcendence within transcendental worlds emerging within Creator's space.
16. It is a transcendental phenomenon of pair of orientations sustaining each other.
17. It is a transcendental phenomenon of pair of half circumferences unifying a complete circumference enveloping a spatial order of the Creator's space.
18. It is a transcendental phenomenon of skylines of transcendental worlds within Creator's space emerging as the sustenance lines of both spatial and solid orders of both Creator's space and of transcendental worlds.

EXERCISES

Chase the transcendental phenomenon of above features, one by one and to evaluate one's comprehension of these features of the transcendental phenomenon.

■

ASPECT 10.5
HEMISPHERE ALONG FIRST AXIS OF SPATIAL ORDER

Exercises about first hemisphere and first axis

1. Chase first hemisphere within Creator's space in terms of first axis of spatial order, firstly as linear dimensions leading to solid domain, secondly as solid domain to be in the role of dimension of a transcendental world.
2. Chase first hemisphere within Creator's space in terms of first axis, firstly availing a manifestation format for it and approaching it as a manifestation layer, and secondly by availing a transcendence range for it approaching it as a transcendence phenomenon happening. ■

Glimpse and be blissful

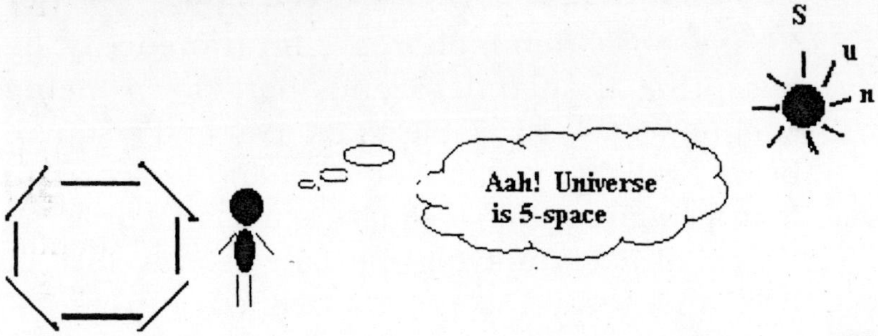

■

ASPECT 10.6
HEMISPHERE ALONG SECOND AXIS OF SPATIAL ORDER

Exercises with focus upon

Second hemisphere and Second axis

1. Chase second hemisphere within Creator's space in terms of second axis of spatial order, firstly as linear dimension leading to solid domain, secondly as solid domain to be in the role of dimension of a transcendental world.
2. Chase second hemisphere within Creator's space in terms of second axis, firstly availing a manifestation format for it and approaching it as a manifestation layer, and secondly by availing a transcendence range for it approaching it as a transcendence phenomenon happening.
3. Chase both hemispheres simultaneously by accepting first and second, both axes, being simultaneously available as interval and a square being available within a square, and to see as to how a square within a square is a spatial order as a spatial axes unifying the pair of universes of first and second axes as a unified universe of spatial dimensional order beings its spatial skyline.

ASPECT 10.7
PARTITION OF UNIVERSE AS HEMISPHERES

Final Exercises

With focus upon partition of universe as a pair of universes within hemispheres

1. Chase partition of '2' as (1, 1).
2. Chase partition of a spatial order as a pair of linear axes.
3. Chase partition of a sphere as a pair of hemispheres.
4. Chase partition of a three dimensional frame into a pair of three dimensional frames of half dimensions.
5. Chase partition of a three dimensional frame of linear units into a pair of three dimensional frame of half dimensions of 'artifice-1' units supplied by half of spatial dimension.
6. Chase a pair of three dimensional frames of 'artifice 1' units constituting a pair of solid dimensional orders within hemispheres.
7. Chase a pair of solid dimensional orders

structuring a pair of transcendental worlds within hemispheres.

8. Chase (1 , 1) as 2.
9. Chase a pair of linear axes as a spatial order.
10. Chase a pair of hemispheres as a sphere.
11. Chase a pair of three dimensional frames of half dimensions as a three dimensional frame of full dimensions.
12. Chase a phenomenon of a pair of transcendental worlds unifying as a single transcendental world.
13. Chase transcendental phenomenon of emergence of transcendental worlds in pairs within Creator's space.
14. Chase Full transcendental world within First hemisphere.
15. Chase Full transcendental world within Second hemisphere.
16. Chase Sustenance of hemispheres by spatial order.
17. Chase Hemisphere along first axis of spatial order.
18. Chase Hemisphere along second axis of spatial order.
19. Chase Partition of Universe as hemispheres.
20. Permit one's transcending mind to glimpse the phenomenon of partition of transcendental universe into a pair of transcendental universes within the Creator's space and also to glimpse the phenomenon of a pair of transcendental universes unifying as a single transcendental world within the Creator's space.

■

FEATURE-11

CHASING DISTINCT UNIVERSES

Last pair of fortnights
Features 9 to 12

Pair of hemispheres

FEATURE-11

ABOUT DISTINCT UNIVERSES

Eleventh feature shall be about glimpsing of the transcendental phenomenon of the skylines partitioning the universe into a pair of hemispheres, on either side of the skylines as distinct pair of universes.

ASPECT-11.1
ARTIFICE SIX AND SIX STEPS LONG RANGES

1. We reached a phase and stage of the course, where we may confront ourselves with a poser as to how we may chase up till the self referral core of the transcendental worlds as seat of self referral domain of 6-space as origin source reservoir of hyper quantifiers for the transcendental worlds.
2. The answer to this poser is to be supplied by artifice '6' and six steps long self referral ranges of the measuring rod, which are to be availed for this chase up till the self referral core of the transcendental worlds.
3. The sathapatya measuring rod is a synthetic set up of representative regular bodies of 1 to 6 space, that is, interval, square, cube and hyper cubes 4, 5, 6, which collectively may be designated as hyper cubes 1 to 6.
4. Lord Vishnu, lord of 6-space is the presiding deity of the measuring rod and Lord Brahma, Lord of 4-space, is the presiding deity of the measure of this measuring rod.

5. The Creator's space / 4-space accepts 0-space as its skyline as it is its dimension of dimension.
6. '0-space' as skyline of Creator's space, makes it a unique space of affine skyline as that 0-space levels all the entities of all the spaces, as $(A^0)^N = 1$ for all A and N.
7. With this, the Creator's space as measures space for the measuring rods makes measures of affine units of same value for all spaces and the measuring rod itself emerges to be a set up as:

   $(A^0)^1 + (A^0)^2 + (A^0)^3 + (A^0)^4 + (A^0)^5 + (A^0)^6$
8. Otherwise, artifice 6 is of unique features, illustratively:
   - (i) 6=1+1+1+1+1+1
   - (ii) 6=1+2+3
   - (iii) 6=1 x 2 x 3
   - (iv) 6=2+2+2
   - (v) 6=3+3
   - (vi) divisors of 6 are 6, 3, 2, 1 with sum value being double of 6 and this makes 6 a perfect number.
9. The artifice 6 as '1, 2, 3', a sequential progression of 3 steps parallel to three axes of 3-space and 3 synthetic units of a measuring rod of 3-space, namely, interval, square and cube, makes it of the features of a solid dimensional order which leads to the transcendental worlds enveloped within Creator's space.
10. This leads to a transition and transformation from 1+2+3=6 to 1+2+3+4=10, and there are six boundary

components of 3-space and 10 boundary components of 5-space.

11. This coordination of 6 boundary components of 3-space and 10 boundary components of 5-space is a coordination of boundaries of 3-space as dimension and 5-space as domain.
12. This coordination is the transition phenomenon of transition from shad chakra format / 6 circuits format of human body to 10 chakra format of human body.

EXERCISES

Express comprehension of

(a) Measuring rod of 6-space

(b) Measure of this measuring rod

(c) 4-space as dimension of 6-space

(d) 4-space supplies measures / units for the measuring rod

(e) The boundary components of hyper cubes 1 to 6 are in the ratio of 2, 4, 6, 8, 10 and 12.

(f) The boundary components of 3-space and 5-space are in the ratio of 6 and 10.

(g) 6 is a perfect number as 1+2+3+6=2 x 6.

(h) Domain boundary ratio of hyper cube 6 is $A^6 : 12B^5$.

■

Sankhiya and Yoga approach

The artifice 6 and hyper cube-6 are of identical formats.

■

ASPECT-11.2
SHAD CHAKAS FORMAT OF HUMAN FRAME: AS EXTERNAL CHARACTERISTICS OF MEASURING ROD

1. The shad chakra / 6 circuits format of human body is described in the scriptures in terms of their external characteristics being 2, 4, 6, 8, 10, 12.
2. This sequential progression is parallel to the boundary components accepted by hyper cubes 1 to 6 and as such these become the external characteristics of the constituents of the measuring rod, that is of hyper cubes 1 to 6.
3. This six steps long circuits format of human body is also parallel to the format of 6 boundary components accepted by cube / 3 space.
4. The artifice 6 is of unique features, first of which is that it is a perfect number as that the sum of its all divisors is double of the value of 6 itself.
5. Further as that 6=1+2+3, it is parallel to the measuring rod of 3-space being constituted by interval, square and cube as representative regular bodies of 1 to 3 space.
6. As such it is of the features of solid dimensional order of the transcendental worlds / 5-space.

7. The transcendental worlds are enveloped within as many as ten Creator's spaces as that the domain boundary ratio for hyper cube 5 is of the formulation $A^5 : 10\ B^4$.

8. The coordination of 3-space and 5-space as dimension and domain and further 3-space accepting 6 boundary component and 5-space accepting 10 boundary component and still further 6=1+2+3 and 10=1+2+3+4 works out a transcendence phenomenon happening through the solid boundary and the same firstly creates intelligence field within Creator's space and then creates consciousness field within the intelligence field.

9. This transcendence phenomenon happening attains transition and transformation for shad chakra format / 6 circuits format of human body into 10 chakras format for human body.

10. Sadkhas fulfilled with intensity of urge to fully know and to completely chase this transcendence phenomenon happening of transition and transformation of human body circuits formats shall permit their transcending mind to glimpse this phenomenon while transcending through the inner layers of the transcendental worlds emerging within Creator's space.

EXERCISES

One shall intellectually validate the experiential glimpsing of the phenomenon of the circuits formats of human body.

■

ASPECT-11.3
TRANSCENDENTAL EXERCISES-1

1. Chase hyper cube 5 for its set up of 5-space enveloped within 4-space.
2. Chase set up of ten hyper solid boundary components of hyper cube 5.
3. Chase the transcendental phenomenon of Lord Brahma, Creator the supreme meditating upon his lord, Lord Shiv and multiply as ten Brahmas.
4. Chase the set up of hyper cube 5 as 3-space playing the role of its dimension and as that 3-space accepts 6 boundary components while 5-space accepts 10 boundary components and thereby there happens transition and transformation from 6 circuits format of human body to 10 circuits format of human body.

■

For to be in a blissful state

One shall permit the mind to transcend time and again to glimpse the inner folds of the transcendental worlds and to get fulfilled with the bliss.

■

ASPECT-11.4
FIVE BASIC ELEMENTS SPACE AND SUN: AS MEASURING ROD

1. There are 5 basic elements namely, Earth, Water, Fire, Air and Space.
2. The sun is ahead of them and these five elements together with Sun constituents a 6 steps long range.
3. Chase this 6 steps long range along the measuring rod of 6-space.
4. With Earth as solid domain / 3-space / and in that sequence being of linear order, water as of spatial order, fire as of solid order, Air as of hyper solid 4 order, space as a hyper solid 5 order and Sun as of hyper solid 6 order, shall be constituting sequential progression of dimensional orders of 5 basic elements and Sun.
5. Parallel to it is sequential progression of domain folds corresponding to the above dimensional orders, viz.,
   (i) Linear order leading to 3-space as domain for earth,

(ii) Spatial order leading to 4-space as domain for water,

(iii) Solid order leading to 5-space as domain for fire,

(iv) Hyper Solid-4 order leading to 6-space as domain for Air,

(v) Hyper Solid-5 order leading to 7-space as domain for Space,

(vi) Hyper Solid-6 order leading to 8-space as domain for Sun,

6. As such the five basic elements and Sun shall be constituting a measuring rod by synthesizing the domain folds of above range of dimensional order that is beginning with linear order and reaching up till hyper solid 6 order.

7. The said range of dimensional orders beginning with linear order and reaching up till hyper solid 6 order shall be constituting the corresponding measures for different steps of above synthetic measuring rod of elements for reaching up till orb of the Sun.

MEASURING ROD

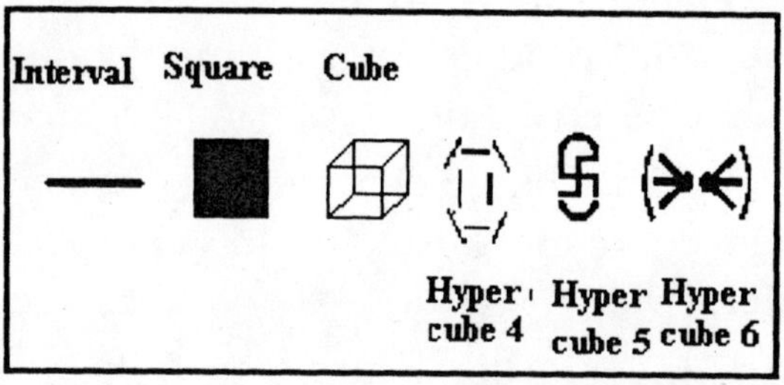

ASPECT-11.5
FIVE BASIC ELEMENTS SPACE AND SUN: AS DIMENSIONAL ORDER

EXERCISES-2

RECAPITULATION

1. It may be recapitulated from the previous day course as that
2. Parallel to it is sequential progression of domain folds corresponding to the above dimensional orders, viz.,

(vii) Linear order leading to 3-space as domain for earth,

(viii) Spatial order leading to 4-space as domain for water,

(ix) Solid order leading to 5-space as domain for fire,

(x) Hyper Solid-4 order leading to 6-space as domain for Air,

(xi) Hyper Solid-5 order leading to 7-space as domain for Space,

(xii) Hyper Solid-6 order leading to 8-space as domain for Sun,

3. As such the five basic elements and Sun shall be constituting a measuring rod by synthesizing the domain folds of above range of dimensional order that is beginning with linear order and reaching up till hyper solid 6 order.
4. The said range of dimensional orders beginning with linear order and reaching up till hyper solid 6 order shall be constituting the corresponding measures for different steps of above synthetic measuring rod of elements for reaching up till orb of the Sun.

EXERCISES

a. Construct a measuring rod in terms of the five basic elements and Sun as domain fold.

b. Chase the measures of different steps of above synthetic measuring rod of elements reaching up till orb of the Sun.

■

For pointed attention

Vykata, Avyakata, ayakato-avyakatat and Pursha

1. Srimad Bhagwad Gita preserves the enlightenment of the order of the Reality as Vyakta, Avyakata, Avyakto, Avyakato-avyakatat and Pursha.
2. Vyakta means expressed. It is the expressed world, Vishwa, 3-space (⊞).
3. Avyakata means unexpressed which is yet to manifest in the Vishwa, 4-space (⌗) world.
4. Avyakato-avyakatat means which is at the base of Avyakata. It is the transcendental base, a 5-sapce world.
5. Pursha is fully expressed transcendental consciousness within hu man frame. It is the consciousness world of *Atman* (soul) domain, *Vishnu-Lok*, the *Pursha* in the *Surya* (sun), the 6-space (⋈) reality.

(From Book Glimpses of Vedic mathematics) ■

ASPECT-11.6
FIVE SENSES BAND AND MIND: AS MEASURING ROD

1. There are 5 senses namely 'touch, taste, smelling, hearing and seeing'.
2. The mind is ahead of them and these five sense together with mind constitutes a 6 steps long range.
3. Chase this 6 steps long range along the measuring rod of 6-space.
4. With these five senses and sixth mind as 'mundane' / domains – contents, shall be constituting 6 steps of the measuring rod.
5. Corresponding to these mundane / domains-contents, there shall be subtle / micro / dimensional level set ups, which shall be constituting a sequential progression range of those micro / dimensional level set ups parallel to the measures for different steps of the senses-mind measuring rod.
6. As such there shall be emerging a pair of sequential progressions ranges parallel to the six steps of measuring rod and there dimensional orders.

7. The senses band and mind, at its mundane level shall be permitting chase along six steps of the measuring rod as has been the case for six basic elements and sun.
8. And further the micro / dimensional level set ups / existence of senses band and mind shall be permitting chase along dimensional orders of six steps of the measuring rod as has been the case for the sequent ional dimensional progression for basic elements and Sun.
9. It would be a blissful experience for the transcending mind to glimpse parallel chase for five basic elements and sun as well as for five senses band and mind along the six steps long measuring rod for mundane / domain level existence as well as for subtle / micro dimensional level existence.

■

For pointed attention

The initiations of young minds are to be specific: (i) about the unison between the *Brahmand* (universe) and the *Pinda* (body), (ii) the potentialities of existence within frame to be of the order of hundred revolution of the earth around the Sun and 120 folds of the creator to envelop the transcendental worlds as transcendental boundary of *Divya Pursha*, (iii) the consciousness state manifesting within the eternal circuit of human frame: (a) *Shad chakras* (six circuits) (b) Nine circuits, (c) ten circuits and (d) multifold inner circuits.

■

ASPECT 11.7
FIVE SENSES BAND AND MIND: AS DIMENSIONAL MEASURES

EXERCISES-3

RECAPITULATION

1. It may be recapitulated from the previous day course as that
2. The senses band and mind, at its mundane level shall be permitting chase along six steps of the measuring rod as has been the case for six basic elements and sun.
3. And further the micro / dimensional level set ups / existence of senses band and mind shall be permitting chase along dimensional orders of six steps of the measuring rod as has been the case for the sequent ional dimensional progression for basic elements and Sun.
4. It would be a blissful experience for the transcending mind to glimpse parallel chase for five basic elements and sun as well as for five senses band and mind along the six steps long measuring rod for mundane / domain level existence as well as for subtle / micro dimensional level existence.

EXERCISES

1. Construct a measuring rod in terms of the five senses band and Mind as domain fold.
2. Chase the measures of different steps of above synthetic measuring rod of senses band reaching up till the mind.
3. Intellectually validate the experience for the transcending mind to glimpse parallel chase for five basic elements and sun as well as for five senses band and mind along the six steps long measuring rod for mundane / domain level existence as well as for subtle / micro dimensional level existence.

■

For pointed attention

Our existence is of many fields wihtin human frame. The first amongst them is the **sensory field** which takes us only up till the spatial boundary of cube.

The next field is the **intelligence field**.Mind creates it during transcendence through the solid boundary of Creator's space (4-space).

Ahead is the **consciousness field**. It emerges within the intelligence field with glimpsing of the transcendental worlds by the transcending mind.

The consciousness field is of different states which the sadkhas attain by glimpsing the inner folds of the transcendental worlds.

■

FEATURE-12
CHASING SELF REFERRALS DOMAINS

Last pair of fortnights
Features 9 to 12

Pair of hemispheres

FEATURE-12

Finally there would emerge as **twelfth feature**, the phase and stage of transition and transformation for the transcending mind glimpsing at core of the transcendental worlds (5-space) the phenomenon of self referral domains (6-space) being lively as the orb of the Sun / Being / Self / Divya Pursha.

ASPECT-12.1
FIVE BASIC ELEMENT RANGE

1. Panch Mahabhut / Five basic elements are five manifestation layers within Creator's space First mahabhut, namely 'Earth' as first basic element, is first manifestation layer, being linear dimensional order, viz. (1-space, 2-space, 3-space, 4-space).
2. If the focus shifts from dimension fold to domain fold, and parallel to it from 'measure' to 'measuring rod', then focus for 'EARTH' would be upon 3-space as 'domain'.
3. Likewise, five basic element shall be permitting depictions as follows:

First element /manifestation layer	1 2 3 4 Earth as domain
Second element / manifestation layer	2 3 4 5 Water as domain
Third element / manifestation layer	3 4 5 6 Fire as domain
Fourth element / manifestation layer	4 5 6 7 Air as domain

Fifth element 5 6 7 8 Space as domain
/ manifestation layer

Now, manifestation of transcendental worlds, shall be bringing into operation, transcendental ranges as well.

And same shall be of features

1 2 3 4 5
2 3 4 5 6
3 4 5 6 7
4 5 6 7 8

First range / quarter

As such, first quarter of elements shall be constituting first transcendental range, and likewise second, third and fourth quarters shall be constituting second, third & fourth transcendental ranges.

The fifth transcendental range is to remain un-manifest base in a SPACE of transcendental worlds.

1 2 3 4 5
2 3 4 5 6
3 4 5 6 7
4 5 6 7 8

5 6 7 8 9

■

ASPECT-12.2
DIVYA PURSHA

'Divya Pursha / eternal existence phenomenon within human frame, is the Reality whose glimpse by the transcending mind would be the attainment of this course exposure of emergence of chain of fields within fields in Creator's space.

It is the attainment phase and stage of course. It is in terms of this attainment that the whole success of being through the course is to be evaluated by each one phenomenon of life within human frame. It would make us enlightenment about 'Divya-pursha', ever lively within experiential range of the transcending mind.

One shall sit comfortably and permit one's mind to transcend to unfold its experiential range and glimpse the transcendental phenomenon of life lively within human frame and marking its presence as virtues emerging within one's consciousness field.

It is this chase of the consciousness field by transcending mind which would be of its order of glimpsing the internal folds of the transcendental worlds. For it, naturally, one is to go in trans time and again and to attempt to glimpse the self-referral core of the transcendental worlds.

It is this phase and stage of glimpsing of the self referral core of the transcendental worlds as their inner most fold which shall be bringing the transcending mind to be face to face with 'Divya-pursha'.

Once one comes face to face with 'Divya-pursha', every thing would get transformed of its own, and rest, and the pilgrimage ahead would go of automation features under the command of 'Divya-pursha'. Transcending mind shall be in a privileged state of being fulfilled with ambrosia of bliss of SELF in command of every thing.

■

To think it out and be blissful

6.3.16 COSMIC ORDER

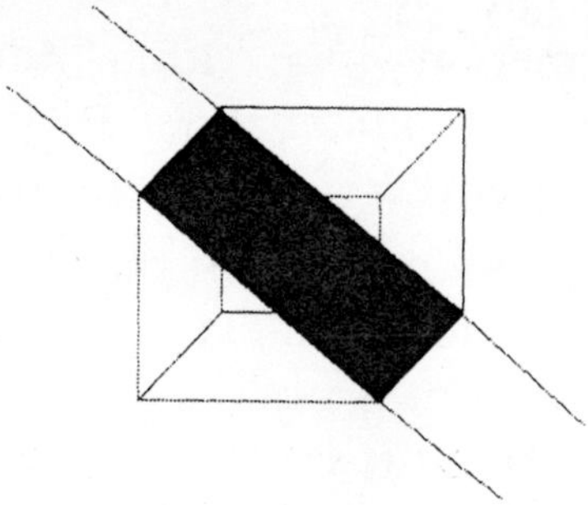

1. NVF (COSMIC) = 62 = 31 + 31 = NVF (CUBE) + NVF (CUBE)
2. NVF (ORDER) = 60 = NVF (FOUR) = NVF (BIBLE) + NVF (BIBLE)

(From Vedic mathematics decodes: SPACE BOOK) ■

ASPECT-12.3
ATMAN (SOUL)

This course, as such brings us to a phase and stage, when our urge to be face to face with one self is going to be fulfilled just with simple attention of the transcending mind to transcend through the —— band range and being there where 'mind' is of its own, lively as SELF.

SELF is lively as 'mind' making its existence as of self-interacting features of spatial order of Creator's space. 'Mind' being spatial order of Creator's space acquires creature powers, with whose coming into play stands created an intelligence field.

With it a process stands set into motion. Intelligence field transforms into a consciousness field. The consciousness field transforms into life field. The life virtues ride the transcendental carriers under the command of the SELF / ATMAN.

It is this command of the SELF which sets the Being on Brahman pilgrimage.

Transcending mind glimpses every thing happening as its creations obeying the commands of the SELF.

Glimpsing it as that 'it itself as well being silently guided by the SELF, goes transcendental ever blissful.

Sadkhas shall continue glimpsing the way the transcending mind transcends through the senses bands, the way it is silently guided by the SELF and the way Being rides the transcendental carriers obeying command of the SELF and sets itself along the Brahman pilgrimage paths.

■

To think it out and be blissful

6.3.14 CREATOR ION AS SOLIDS PAIRING

1. NVF (CREATOR ION) = 118 = NVF (SOLID, SOLID).
2. NVF (SOLID) = NVF (LINEAR); and as such, the outer parts of tri-monad with emergence of Ion field spatial order for the central part, transformed from "Linear" to "solid".

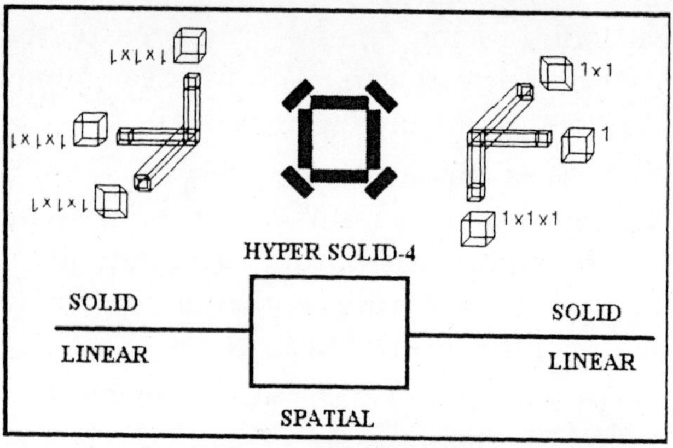

(From the Vedic mathematics decodes: Space Book)

■

ASPECT-12.4
VISHNU LOK

Creator's space, transcendental worlds and self referral domains are the three distinct domains / worlds ahead of our Triloki.

Triloki, Creator's space, transcendental worlds and self referral domain are four quarters of Reality with whose transcendental takes to the Sapt rishi lok / unity state pole star universe.

Triloki (3-space) as dimension fold / first quarter, Creator's space (4-space) as boundary fold / second quarter, transcendental worlds (5-space) as domain fold / third quarter and Vishnu lok (6-space) as origin fold / fourth quarter is essentially the transcendental layer of the Reality.

Vishu lok being the origin resource reservoir of the transcendental layer, as such, it is of unique features and because of it is of the order of God (sixth) state of consciousness. It accepts 26 basic elements and accordingly is coverage in terms of 26 meters / chandas; each of which being of four quarter. This range of 26 elements, on its pairing leads to 52 letters vedic alphabet.

Twenty six elements as 26 artifices of numbers 1 to 26, together as an alphabet of 26 letters A to Z shall be leading to number value formats for words (formulation) EARTH and six being equal, as 52.

This, on its chase, shall be melting our mental blocks as that English alphabet, and as a matter of fact all orthodox and classical alphabet are just the mutations of the Vedic alphabet Need only is that one shall melt one's mental block by transcending the transcendental layer itself and attaining unity state lively ahead.

One shall continue transcending and let the transcending mind glimpse the transcendental phenomenon of unity state and also the automation domains of NATURE (Asht-prakriti (8-space) and the ultimate affine state of nav-Brahman (9-space) transcendently over lively for the transcending mind.

To Glimpse and be blissful

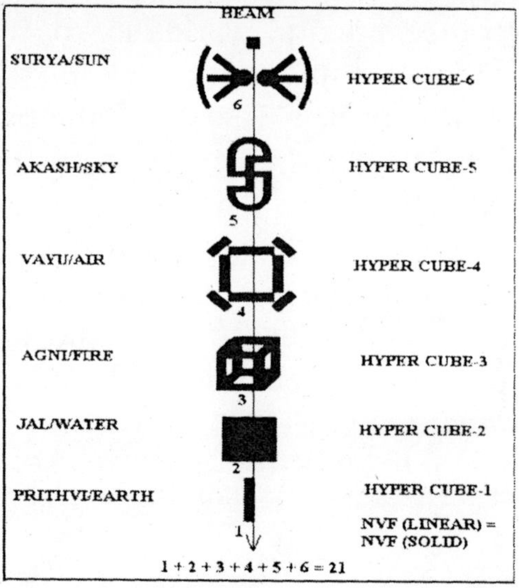

ASPECT-12.5
SUN

Reality chase takes us from 'earth' to 'sun'. It also may be chased from 'Sun' to 'earth'. 'SUN' to 'EARTH' is a six steps long range. This range may be taken parallel to the Sathapatya measuring rod with Lord Brahman as the presiding deity of the 'measure' and 'Lord Vishnu' as the presiding deity of the 'measuring rod'.

'Orb of the Sun' is to be chased as 6-space domain. Its 'origin' is to be the transcendental core of unity state of the order of the Pole-Star'.

'Sun' is the in between state. For Brahman pilgrimage, the take off is to take place from within the orb of the Sun through transcendence at its transcendental core of unity state being of the order of Pole-star / Sapt rishi lok / 7-space.

With 3-space in the role of dimension leads us to 6-space as origin fold of transcendental / 5th manifestation layer (3, 4, 5, 6).

With 6-space in the rule of dimension leads us to 9-space as origin fold of Brahman / 9th manifestation layer.

The synthesise range (3, 4, 5, 6) and (6, 7, 8, 9) with dual role for 6-space as of origin fold as well as of dimension fold is to be of 7 steps long, namely (3, 4, 5, 6, 7, 8, 9).

The features of this range are of measuring rod of unity state which as measure range (1, 2, 3, 4, 5, 6, 7) works and measuring rod being (3, 4, 5, 6, 7, 8, 9). It is this feature which sharpens edges of a cube to be 'lives', which in fact are constituted by solid-points.

Another feature of this measuring rod is that, at its ends, when beginning is as (3^1), the end would be (3^2). And middle is (3^1+3^1).

Then, 6 = 1+2+3 = 1 x 2 x 3 = 2+2+2 And, for all these features, six-space, is to be of spatial as well as solid skylines.

And this way Sun transforms as twelve transcendental Suns / Adityas, a transcendental phenomenon, which one shall go on glimpsing, time and again, and to be blissful.

■

To glimpse and be blissful

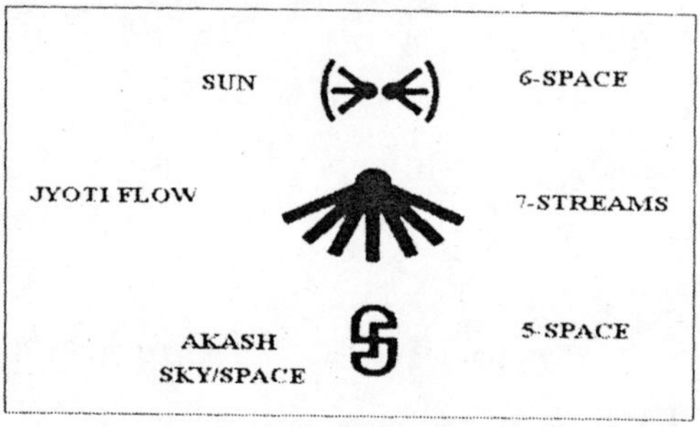

■

ASPECT-12.6
HYPER CUBE 6

New we reach a phase and stage of culmination state of emergence of life-field as of hyper cube 6 format, frame, flow and formations within a Creator's space.

By transcending through the solid boundary, the transcending mind enters the Creator's space. Simultaneously the transcending mind creates its own intelligence field as of hyper cube 5 format, frame, flow and formations within Creator's space.

By continuous transcending, Sadkhas shall be perfecting their intelligence.

To be through solid-boundary, is only one of the roles of Solids / cubes / 3-space. The other roles of solids are as of solid dimensional which shall leading to 6-space as origin fold, and of 3-space itself as origin fold for which the dimensional order would of 0-space order which otherwise is the skyline of 6-space.

As such, from 6-space as origin fold to 0-space as its skyline and '0-space' as dimensional order to lead up till 3-space as original fold which in dimensional role to lead to 6-space as origin fold. This happening and coordination within hyper cube 6 as origin fold sets into motion a process of the emergence of intelligence field, its perfection states, transforming into consciousness state which further on their unity state transforming further into life-field.

Intelligence waves carry the consciousness impulses, the consciousness impulses carry the life virtues and the life virtues riding the transcendental carriers within rays of the Sun reach orb of the Sun availing format of hyper cube 6 like that the transcending mind glimpses the transcendental phenomenon of sequential emergence of intelligence within 4-space, consciousness field within intelligence field and life field within consciousness field. And life virtues riding the transcendental carriers within rays of the Sun and reaching orb of the Sun. From where takes off the being for onward Brahman pilgrimage.

One shall continue transcending and have privilege of glimpsing the Being on Brahman pilgrimage.

■

For focused attention

The symbolic expression of hyper cube-6, as under is for focused attention.

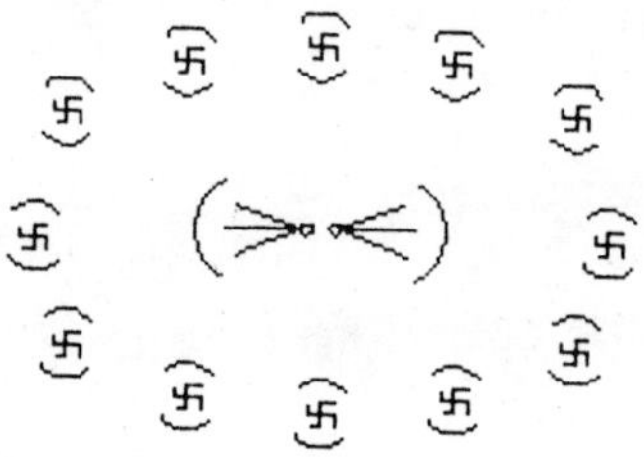

■

ASPECT-12.7
SIX SPACE

Here we reach the concluding day of the course. The topic of the day as well is the all en-compassing 6-space.

Six-space is all en compassing. It is of the order of Vishnu lok. It is the orb of the Sun. It is atman / Self. 'Divya-pursha' is lively within orb of the Sun. It supplies format of its representative regular body, viz. of hyper cube 6, for manifestation of human body. It is, so designated and is known pursha format.

The shed chakra format of human frame runs parallel to the measuring rod of six-space.

Creator's space plays the role of dimension of Vishnu lok. This is parallel to Sathapatya acceptance of Lord Brahman, creator the Supreme, being the presiding deity of 'Measure' of the measuring rod, while Lord Vishnu himself being the presiding deity of the measuring rod.

The transcendental phenomenon of emergence of transcendental worlds within the Creator's space with Creator's space getting fulfilled with solid quantifiers because of the Solid order of the origin source reservoir of Creator's space.

With transcendental happening of emergence of transcendental worlds within Creator's space makes the measure of the measuring rod being of transcendental features and with it the Vishnu lok gets enveloped within the transcendental boundary and the 'unity core' as origin resource reservoir of Vishnu lok ensures transcendental order supply within Vishnu lok and thereby, inward and outward, everywhere comes into play of transcendental features.

The boundary is transcendental. The transcendental quantifiers are fulfilled in Vishnu lok. The unity core of the Vishnu lok is of transcendental order, of as many as seven transcendental dimensions.

With this transcendence taking place at the boundary, within domain as well as within origin fold and also within all the seven transcendental dimensions.

Skylines emerging at the boundary within domain, within origin fold at unity core and also within each of the transcendental dimensions.

With it, let us be grateful to the transcendental bliss being there all-round for the transcending mind to glimpse.

■

To glimpse further

To glimpse the core of the transcendental worlds as seat of self referral domain of Vishnu lok / 6-space / hyper cube-6 accepting as many as 13 versions.

■

VEDIC GEOMETRY COURSE

This course has been made available from 15-9-05 to 7-12-05. The lessons 1 to 84 have been regularly e-mailed to the participants on daily basis. Shri Rohtash Kapoor deserves appreciation for providing this service.

FIRST FEATURE COURSE SCHEDULE INTIMATION

We take it a privilege to share the schedule of first feature of the course as under. The expectation of the course is that the participant is regular and conscious of the steps. One shall think and mediate about the different aspects of the feature under focus during the week. Further expectation of the course is that participant shall daily EXPRESS ONESELF, the way and extent, one could comprehend the aspects of the FEATURE. If these 'expressions' would be share with us, these would provide us an opportunity to help you to evaluate yourself about your comprehension. One shall preserve one's attempts / expressions, as these ultimately will help chase one's sequential progress of comprehension chase. This also will help have an insight as to the different phases and stages of mental state through which one has to transcend to reach at the comprehension of the first feature of Vedic Geometry which is about the 4-space Reality.

This course had begun on 15th Sepember, 2005 and the interested sadkhas were welcomed for this free course with the following welcome letter.

"Welcome-letter"

"Dear Sadkhas

It is our privilege to welcome you to the Course.

At the outset, it is shared with you that Kenneth Williams and R. P. Cosmic Kapoor have very kindly agreed to pool in their knowledge and experience for the sadkhas to successfully achieve the aim of this course.This is an initial stage course. It aims to take up only twelve basic features of the Discipline of Vedic Geometry. Object of the course is to put you on interactive participation for the chase of the twelve basic concept, of which one is to be taken up every week during these three months.

It is being shared at the outset, as that, like all other disciplines of knowledge, the Vedic Geometry as well has reached us with its identity and results merged within the common resource of whole range of knowledge as a single discipline of Vedic knowledge.

As such, the Discipline of Vedic geometry, at it initial stage, is to be chased in terms of the basic features of the Discipline which are over flowing the Vedic knowledge with fervor.

Of these, the most basic feature is the Creator's space presided over by the creator, the supreme, Lord Braham, the four head Lord. The characteristics of IDOL of Lord Brahma, accordingly emerged to be the first resource pool of values of this feature of Vedic geometry focusing upon the Creator's space as Real 4-space with format of its representative regular body manifesting as an IDOL of the Creator.

Scriptures preserve the characteristics features of IDOL of Lord Brahma, Creator the Supreme, the four head Lord with a pair of eyes equipped in each head, as sitting majestically on a lotus seat of eight petals and meditating gracefully within cavity of His Heart upon His Lord, the Lord of transcendental worlds, Lord Shiv, the Five head Lord with three eyes equipped in His each Head.

As such, here, formally we take up the first feature of the course.

■ Dr. S. K. Cosmic Kapoor"